Lecture Notes
in Control and Information Sciences 395

Heidar A. Talebi, Farzaneh Abdollahi,
Rajni V. Patel, Khashayar Khorasani

Neural Network-Based State Estimation of Nonlinear Systems

Application to Fault Detection and Isolation

Authors
Heidar A. Talebi
Department of Electrical Engineering
Amirkabir University of Technology
424 Hafez Ave.
15914 Tehran
Iran
alit@aut.ac.ir

Farzaneh Abdollahi
Department of Electrical Engineering
Amirkabir University of Technology
424 Hafez Ave.
15914 Tehran
Iran
f_abdoll@ece.concordia.ca

Rajni V. Patel
Department of Electrical & Computer Engineering
University of Western Ontario
1151 Richmond Street North
London ON N6A 5B9
Canada
rajni@eng.uwo.ca

Khashayar Khorasani
Department of Electrical & Computer Engineering
Concordia University
1455 Maisonneuve Blvd.
West, EV005.126
Montreal QC H3G 1M8
Canada
kash@ece.concordia.ca

ISSN 0170-8643 e-ISSN 1610-7411
ISBN 978-1-4419-1437-8 e-ISBN 978-1-4419-1438-5
DOI 10.1007/978-1-4419-1438-5
Springer New York Dordrecht Heidelberg London

Library of Congress Control Number: 2009940450

Printed on acid-free paper

Springer is part of Springer Science+Business Media (www.springer.com)

To

My Family (H. A. T.)
My Parents (F. A.)
My Family (R. V. P.)
My Family (K. K.)

Preface

The state of a process specifies its behavior, and many control schemes such as inverse dynamics and feedback linearization rely on the availability of all the system states. However, in many practical systems only the input and output of a system are measurable. Therefore, estimating the states of a system plays a crucial role in monitoring the process, detecting and diagnosing of faults, and achieving better performance. Furthermore, most practical systems are nonlinear, and using linearization or quasi-linearization methods limits the estimation accuracy to a small dynamic range. Several conventional nonlinear observers have been proposed during the past couple of decades. However, most of this work relies on exact *a priori* knowledge of the system nonlinearities. This assumption is rarely satisfied for most practical processes where determining an exact model is quite a difficult, if not impossible, task. Robot manipulators with flexible joints or links are good examples of such systems. Flexibility in a manipulator causes extreme difficulty in modeling manipulator dynamics and becomes a potential source of uncertainty that can degrade the performance of the manipulator and in some cases can even destabilize the system. Thus, model-based observers are not best suited for such systems.

Capabilities of neural networks for identification, observation and control of nonlinear systems have been investigated in both off-line and online environments. In fact, the adaptive behavior of neural networks makes them powerful tools for state observation without any *a priori* knowledge about the system dynamics. Several neural network-based observers have been proposed in the literature for state estimation of nonlinear systems. However, most of these techniques suffer from restrictive assumptions such as (a) a strictly positive real (SPR) condition on the output error equation, (b) scalar-valued nonlinear functions, (c) Linear-in-Parameter Neural Networks (LPNN), (d) a special class of nonlinear systems (e.g. affine nonlinear systems), (e) lack of proof of stability, and (f) a complex weight-updating mechanism, which prevent the use of such observers to real-world applications.

On the other hand, in many control applications unpredictable behavior such as poor performance or even unsafe operation can result from small abnormal deviations (malfunctions) either in the sensors and actuators, or in the components of the process. Hence, an exceptional level of autonomy is required. Recognizing that fault detection and identification is an essential capability of an autonomous system, there is a high demand for development of novel methods for fault detection, isolation, and recovery systems.

The objective of this monograph is to address the problem of state estimation, system identification and observer-based fault detection and isolation (FDI) for nonlinear systems. Towards this end, a neural network-based adaptive observer for a general model of MIMO nonlinear systems is first proposed with no *a priori* knowledge about the system nonlinearities. The neural network is *nonlinear* in its parameters and can be applied to many systems with arbitrary degrees of nonlinearity and complexity. The online weight-updating mechanism is a modified version of the backpropagation algorithm with a simple structure together with an e-modification term that is added for enhanced robustness to unmodelled dynamics and uncertainties. The SPR assumption imposed on the output error equation is also relaxed. The proposed structure is then employed for the system identification problem. The proposed state estimation scheme is employed to develop a new observer-based fault detection and isolation scheme. Several types of faults, namely actuator bias faults, actuator gain faults, and sensor bias faults are considered. The proposed method relies on only output measurements and is also robust to dynamic uncertainties as well as disturbances and measurement noise. Moreover, the fault detection, isolation, and estimation steps are all unified, i.e., neither extra measured/calculated signals nor a separate fault isolation policy is required to isolate the faults. For each developed algorithm, mathematical proofs of stability are given by using Lyapunov's direct method. The effectiveness of our proposed state estimation/identification/fault detection approaches is demonstrated through extensive simulations as well as experimentations that are carried out on highly nonlinear systems. The case studies include flexible-joint and flexible-link manipulators, satellite attitude control systems with reaction wheel and magnetorquer type of actuators (simulations), and a 3 DOF macro-micro manipulator and a 6 DOF industrial manipulator, namely the PUMA 560 (experiments).

The funding for much of the research described in this monograph was provided by the Natural Sciences and Engineering Research Council (NSERC) of Canada (Patel and Khorasani), by a Tier-I Canada Research Chair (Patel), by an infrastructure grant from the Canada Foundation for Innovation awarded to the University of Western Ontario (Patel), by the Faculty of Engineering and Computer Science (Khorasani), and Tier-I Concordia University Research Chair (Khorasani).

June 2009,

H.A. Talebi
F. Abdollahi
R.V. Patel
K. Khorasani

Contents

List of Figures

List of Tables

Acronyms

ACS	Attitude Control Subsystems
BP	Backpropagation
EKF	Extended Kalman Filter
FDI	Fault Detection and Isolation
FSMS	Flexible Structure Mounted manipulator Systems
LEO	Low-Earth Orbit
LPNN	Linear-in-Parameters Neural Network
M^3	Macro-Micro Manipulator
MIMO	Multi-Input Multi-Output
NLPNN	Nonlinear-in-Parameters Neural Network
NN	Neural Network
PR	Positive Real
RBF	Radial Basis Function
SISO	Single-Input Single-Output
SPDM	Special Purpose Dexterous Manipulator
SPR	Strictly Positive Real

Chapter 1
Introduction

In this chapter, several important concepts concerning design and implementation of neural network observers are introduced. The motivation and objectives of the monograph are given in Section 1.1. In Section 1.2, background material on multilayer neural networks and structure of our proposed neural network observers are presented. Finally, the outline of the monograph is provided in Section 1.3.

1.1 Preamble

The state of a process specifies its behavior and many control schemes such as inverse dynamics and feedback linearization rely on the availability of system states for analysis and implementation. However, in many practical systems only the inputs and outputs of the plant rather than the full state vector can be measured as use of measurement tools may not be feasible or economical. In such cases, unmeasured states have to be inferred from a measurable subset of the states. For instance, in process monitoring of biological systems [9], some process variables cannot be measured directly, e.g. concentration in a complex medium with solid substrates. Moreover, accuracy of most online measurement systems can be low. Therefore, biomass, a key variable, has to be evaluated from other measurable variables by appropriate state estimator. In a chemical process, effective control and monitoring of the process relies on real-time information about the state variables of the process. A reliable online information on unmeasurable state variables should be obtained by using a state estimator. For example, in chemical reactor the temperature of the reactor inlet and outlet streams are used to estimate the extent of reactant conversion in the reactor [10].

On the other hand, the main objective of an automatic control system, i.e., assuring acceptable performance cannot be achieved only by improving the individual radiabilities of the functional units. It also requires an efficient fault detection, isolation and accommodation capability. In the control systems community, a fault is understood as any kind of malfunction in the actual system which leads to an

H.A. Talebi et al., *Neural Network-Based State Estimation of Nonlinear Systems*, Lecture Notes in Control and Information Sciences 395,
DOI 10.1007/978-1-4419-1438-5_1,

anomaly in the overall system performance. Such malfunctions may occur either in the sensors (instruments), actuators, or in the components of the process. The majority of fault detection and isolation approaches (e.g. those based on analytical and/or knowledge redundancy) require advanced information processing techniques.

In fact, the basic idea of analytical redundancy deals with checking the actual system behavior for consistency with a nominal mathematical model. A number of methods have been introduced in the literature for analytical redundancy based on basic concepts such as: (a) parity space approaches [11, 12], (b) dedicated observer and innovation based approaches, (c) fault detection filter approaches [13], and (d) parameter identification approaches [14]. For the first three approaches, a proper state estimator is required. Moreover, parameter identification can also be considered as a simplified problem of designing a state estimator [15, 14]. Therefore, estimating the states of the system plays a crucial role in monitoring the process [9, 10], detecting and diagnosing faults [16, 17, 18, 19, 20], and achieving desired performance requirements.

Several conventional linear as well as nonlinear observers have been suggested during the past decades some of which are listed as follows:

- In 1964, Luenberger introduced his well-known state estimator for linear systems [21]. The Luenberger observer is indeed the simplest type of the closed-loop observers. In this observer, the output error will be fed back to the observer. This gain is selected such that $(A-LC,C)$ be observable.
  Consider the following state space representation:

$$\begin{aligned} \dot{x}(t) &= Ax(t) + Bu(t) \\ y(t) &= Cx(t). \end{aligned} \tag{1.1}$$

  Then the structure of the Luenberger observer is given by

$$\begin{aligned} \dot{\hat{x}}(t) &= A\hat{x}(t) + Bu(t) + L(y(t) - \hat{y}(t)) \\ \hat{y}(t) &= C\hat{x}(t). \end{aligned} \tag{1.2}$$

  This observer can also be used for the plant with deterministic noises.
- **The Kalman filter** is one of the most popular state estimators for linear systems which are affected by noise. In fact, Swerling proposed a solution for optimal estimation of orbits and trajectories of satellites and missiles in [22], which was later refined by Kalman and is now known as the Kalman filter [23]. This observer can be applied to the plant with stochastic noises in states and the measurements [24]. The following model is considered for the plant

$$\begin{aligned} \dot{x}(t) &= Ax(t) + B_u u(t) + B_w w(t) \\ y(t) &= Cx(t) + v(t), \end{aligned} \tag{1.3}$$

  where $w(t)$ and $v(t)$ are uncorrelated white noise with the spectral densities S_w and S_v, respectively;

$$E[w(t)w^T(t+\tau)] = S_w\delta(\tau)$$
$$E[v(t)v^T(t+\tau)] = S_v\delta(\tau).$$

Then, the observer is expressed as

$$\begin{aligned}\dot{\hat{x}}(t) &= A\hat{x}(t) + B_u u(t) + G(t)(y(t) - \hat{y}(t)) \\ \hat{y}(t) &= C\hat{x}(t).\end{aligned} \tag{1.4}$$

The Kalman filter gain $G(t)$ can be obtained in terms of the estimation error covariance matrix

$$G(t) = \Sigma_e(t)C^T S_v^{-1},$$

where

$$\Sigma_e(t) = E([x(t) - \hat{x}(t)][x(t) - \hat{x}(t)]^T).$$

It is clear that, the structure of the Kalman filter is similar to that of the Luenberger observer except that its gain $G(t)$ varies in time.
By linearizing the nonlinear dynamics about the current mean and covariance, this filter is modified as the extended Kalman filter (EKF) for application to estimating the state of nonlinear systems. However, unlike its linear counterpart, the extended Kalman filter in general is not an optimal estimator [25, 26]. If the initial estimate of the state is not properly chosen, or if the process is not precisely modeled, the filter may quickly diverge.

- **High-gain observer** which takes advantage of the high-gain as a classical tool for compensating nonlinearities is proposed in [27]. It is widely used for estimating the states of complicated systems in the literature [28, 29, 30, 31, 4]. This strategy can be applied to systems which are uniformly observable for any u(t), i.e., the states of the system can be determined from the output of the system and its derivatives independent of the input. In other words, the plant should have the following structure

$$\begin{aligned}\dot{x}_1 &= x_2 \\ \dot{x}_2 &= x_3 \\ &\vdots \\ \dot{x}_{n-1} &= x_n \\ \dot{x}_n &= \varphi(x, u) \\ y &= x_1,\end{aligned} \tag{1.5}$$

where $x = [x_1, ..., x_n]^T$ and $u = [u, \dot{u}, ..., u^{(n)}]^T$. The structure of the high-gain observer is a simple chain of integrators such that each of them is corrected by the injection of the output error $(y - \hat{y})$ multiplied by a factor depending on the observer gain k. The structure of the high-gain observer is given below:

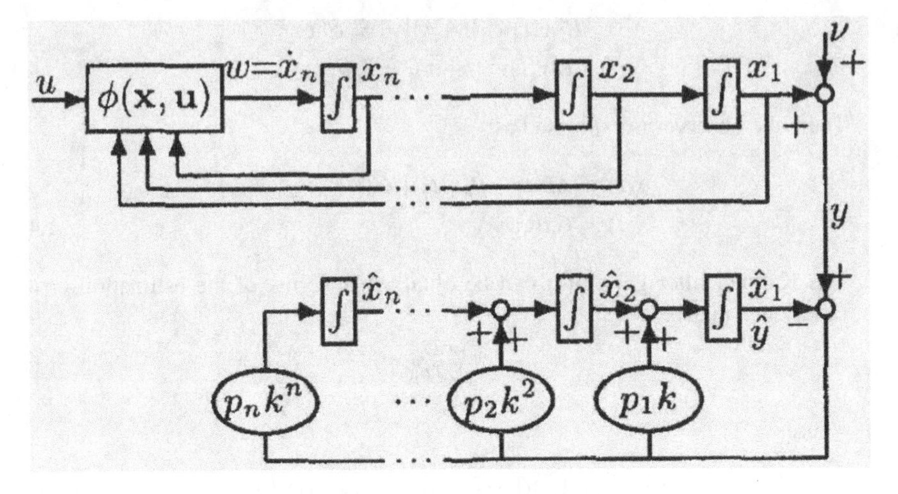

Fig. 1.1 High-gain observer for a system in observability normal form [4].

$$
\begin{aligned}
\dot{\hat{x}}_1 &= \hat{x}_2 + p_1 k(y-\hat{y}) \\
\dot{\hat{x}}_2 &= \hat{x}_3 + p_2 k^2(y-\hat{y}) \\
&\vdots \\
\dot{\hat{x}}_{n-1} &= \hat{x}_n + p_{n-1} k^{n-1}(y-\hat{y}) \\
\dot{\hat{x}}_n &= p_n k^n(y-\hat{y}) \\
\hat{y} &= \hat{x}_1,
\end{aligned}
\tag{1.6}
$$

where $\hat{x} = [\hat{x}_1, ..., \hat{x}_n]^T$ denotes the estimation of the states x. In contrast to the classical observers such as Luenberger observer and Kalman filter, the high-gain observer does not consist of a replica of the system plus correction terms as the nonlinearity $\varphi(x,u)$ is not modeled. Figure 1.1 shows the schematic of the high-gain observer. The observer error will be defined as $e_i(t) = x_i(t) - \hat{x}_i(t)$. Now, by simple calculation, one can get

$$
e_1(s) = \frac{1}{(s+k)^n} w(s) + \frac{s^n}{(s+k)^n} v(s), \tag{1.7}
$$

where $w(s) = L\{\dot{x}_n(t)\}$, $L(.)$ denotes the Laplace transform and $v(s)$ is the additive output noise. It is obvious that the larger observer gain k is chosen, the smaller influence of w and thus of the nonlinearity φ (and therefore of u) on the observer output error. By selecting the observer gain large enough, the system will have excellent robustness properties (the name "High-Gain"). However, for large values of k the additive output noise is damped in the output error and therefore undamped in the observer output. Consequently, in practical applications the determination of an appropriate value for the observer gain is a major issue. Several algorithms are proposed for this purpose [29, 4]. A simple adaptive algorithm proposed in [4] is given below:

$$\frac{d}{dt}s(t) = \begin{cases} \gamma|y(t)-\hat{y}(t)|^2 & \text{for } |y(t)-\hat{y}(t)| > \lambda \\ 0 & \text{for } |y(t)-\hat{y}(t)| \leq \lambda \end{cases}$$
$$t_i : s(t_i) = S_i,\ i = 0,1,2,...$$
$$k(t) = S_i,\ \forall t \in [t_i, t_{i+1}), \tag{1.8}$$

where $\lambda > 0,\ \gamma > 0,\ \beta > 0, S_0$ are given and

$$S_{i+1} - S_i = \beta e^{i^2}\ \ \forall \geq 0. \tag{1.9}$$

The idea behind this adaptation law is that the observer gain k is piecewise constant and takes values S_i in each interval given in (1.8). The switching time for the new S_i depends on the monotonically increasing parameter s. Whenever s reaches a new threshold S_i (the time when this occurs is denoted by t_i), the observer gain takes the value of S_i. The S_i are predefined as a monotonically increasing sequence such that their growth rate is larger than e^i. One possibility to guarantee this is given in (1.9). Thus $k(t)$ is increased step-wise as long as $y - \hat{y}$ lies outside λ and cannot be decreased and p is defined as $p_i = \begin{pmatrix} n \\ i \end{pmatrix}$. Although, the exact knowledge about the dynamics of the plant is not required for this algorithm, the system dynamics must be in a special form. Moreover, the system output cannot have unmeasured disturbance.

- **Sliding mode observer** provides robust performance for observing the states of nonlinear systems with uncertainties [32, 33, 34, 35].
  It is well known that modeling inaccuracies can have strong adverse effects on nonlinear observer systems. Therefore, any practical design must address them explicitly. Two major and complementary approaches to dealing with model uncertainty are robust and adaptive approaches [32]. In fact, the sliding mode observer was shown successful in dealing with model-free systems. The idea is to define a time-varying surface ($S(x,t)$) in the state-space such that the problem of $x = \hat{x}$ is equivalent to that of remaining on the surface $S(t)$ for all $t > 0$. The major difficulty here is how to find a suitable sliding surface ($S(t)$).
  To design a sliding mode observer,let us consider the general state space model

$$\dot{x} = f(x,u) \tag{1.10}$$
$$y = Cx,$$

where $x \in R^n$ is the state vector, $u \in R^m$ is the bounded control input, and C is known. To perform a stable integration, a Hurwitz matrix A is defined such that the pair of (A,C) is observable. Then by adding and subtracting Ax, (1.10) modifies as

$$\dot{x} = Ax + F(x,u)$$
$$y = Cx, \tag{1.11}$$

where $F(x,u) = f(x,u) - Ax$. The sliding mode observer is then given by

$$\dot{\hat{x}} = A\hat{x} + S(\hat{x},u) - Ke$$
$$\hat{y} = C\hat{x}, \tag{1.12}$$

where e is the output error $y - \hat{y}$.
In the sequel, the method given in [35] for defining S is briefly presented:

$$S(\hat{x},u) = -\rho p^{-1} sign(e) = \frac{-p^{-1}C^T C\delta}{||C\tilde{x}||}, \tag{1.13}$$

where ρ is a positive constant. Introducing the observer error as $\tilde{x} = x - \hat{x}$, the error dynamics can be written as

$$\dot{\tilde{x}} = -A_0\tilde{x} + S(\hat{x},u) - F(x,u), \tag{1.14}$$

where $A_0 = A - KC$. Since (A,C) is observable, there exists a K such that A_0 is stable. Consequently, the following Lyapunov equation has a positive definite solution

$$A_0^T p + pA_0 = -Q, \quad Q = Q^T > 0. \tag{1.15}$$

If the nonlinear function $f(x,u)$ satisfies the following boundness condition

$$f(x,u) - Ax = -p^{-1}C^T h(x,u),$$

where $\| h(x,u)\| < \rho, \rho > 0$, then by considering the Lyapunov candidate $V = \tilde{x}^T p\tilde{x}$ and differentiating it along the (1.14), the negative definiteness of $\dot{V}$ and consequently stability of the observer will be guaranteed.

It should be noted that some of the introduced observers rely on exact *a priori* knowledge of the system nonlinearities. On the other hand, several nonlinear observers have been suggested for special systems. For instance, [36, 37] provide observers for second order mechanical systems. Whereas, for most practical process, defining an exact model is a hard task or is not simply possible. Robot manipulators with flexible joints or links are good examples of such systems. Flexibility in a joint causes extreme difficulty in modeling manipulator dynamics and becomes a potential source of uncertainty that can degrade the performance of the manipulator and in some cases can even destabilize the system [38]. Thus, model-based observers are not best suited for such systems.

Capabilities of neural networks for identification, observation and control of nonlinear systems have been investigated extensively both off-line and online environments. In fact, adaptive behavior of neural networks make them powerful tools for performing state observation without *a priori* knowledge of the system dynamics.

In this monograph, a stable neural network-based observer scheme is presented for nonlinear systems when there is no *a priori* knowledge about the system dynamics. The main advantage of our proposed observer is that the proposed learning rules are based on a modified backpropagation algorithm and therefore, it can be easily implemented. Moreover, stability of the overall system and the boundedness of the estimation errors are formally guaranteed. Since a nonlinear system identifier can be considered as a class of nonlinear observer with no *a priori* knowledge about the system dynamics when all the states are available, our proposed neural network is applied for system identification.

The above results are subsequently employed for designing observer-based actuator as well as sensor fault detection and isolation schemes. The performance of our proposed observers, identifiers, and FDI schemes are evaluated by both simulations and experimentations on flexible link/joint manipulators as well as the satellite's attitude control subsystems (ACS) with reaction wheel and magnetorquer type actuators.

1.2 Background

In this section, structures of multilayer neural networks and their specifications are reviewed. An overview of techniques developed in the literature on neural network-based observers are also presented.

1.2.1 Multilayer Neural Networks

Artificial neural networks (NNs) which are inspired by biological nervous systems, are promising and powerful tools in identification and modeling of nonlinear systems due to their impressive merit in data processing and learning capabilities. As the name Multilayer Neural Networks suggests, the network consists of multiple layers of neurons which are usually considered as three layers, namely input layer, hidden layer and output layer. However, the number of neurons in each layer is to be determined and can be defined based on the designer choice. Fig. 1.3 shows a three layer neural network where the two weighting matrices, W and V are depicted by arrows. At each neuron of hidden and the output layers, an activation function is defined that is responsible for determining the nonlinear characteristics of the neural network. The output layer activation functions are usually considered as linear functions and typical choices of the activation functions in the hidden layers are hard limit, linear threshold, sigmoidal, hyperbolic tangent, and Gaussian functions as shown in Fig 1.2.

The real capability of multilayer NNs is due to their ability in approximating nonlinear functions as reported in the literature [39, 40, 41, 36, 26] (and the references therein).

The Cybenko theorem [39] shows that a feedforward neural network is capable of approximating any continuous, multivariate function to any desired degree of accuracy. The limitation of any neural network to map a function arises from poor selection of the neural network parameters and weights or an insufficient number of the hidden neurons. The formal statement of this theorem is given below:

Theorem 1.1. *[39] Let I_n denote the n-dimensional unite cube, i.e., $[0,1]^n$, and also the space of continuous functions on I_n is represented by $C(I_n)$. By considering a continuous function $\sigma(.)$ which satisfies the following specification*

$$\sigma(t) \rightarrow \begin{cases} 1 \; as t \rightarrow +\infty \\ 0 \; as t \rightarrow -\infty \end{cases},$$

given any function $f \in C(I_n)$ and $\varepsilon > 0$, there exist vectors W_j, V_j and b_j for $j = 1,...,N$ and a function $G(x)$ such that

$$|G(x) - f(x)| < \varepsilon \; for \; all \; x \in I_n,$$

where

$$G(x) = \sum_{j=1}^{N} W_j^T \sigma(V_j^T x + b_j) = W^T \sigma(V^T x + b), \tag{1.16}$$

$W = vec\{W_j\}$, $V = vec\{V_j\}$, and $b = vec\{b_j\}$.

To model, predict, or identify unknown system dynamics, NNs should be trained to capture the desired systems mapping and functions. Indeed, NNs learn by modifying their weights to reach optimum values. One of the Achilles's heel of designing a NN is developing a proper learning rule. Several techniques for updating rules having different specifications are proposed in the literature some of which are discussed in [42]. A common and popular weight-tuning algorithm is the *gradient algorithm* based on the *backpropagated error* [43] which has found ample of applications in classification, pattern recognition, internal representation, encoding, system identification and modeling [42] (an the references therein).

The *universal approximation theorem* stated in [41] shows that a three layers NN with a backpropagation training algorithm has the potential of behaving as a universal approximator. The *universal approximation theorem* is formally described as follows:

Theorem 1.2. *[41] Given any $\varepsilon > 0$ and any $\mathcal{L}_2$ function $f : [0,1]^n \in \mathcal{R}^n \rightarrow \mathcal{R}^m$, there exists a three-layer backpropagation network that can approximate f within ε mean-square error accuracy.*

Fig. 1.4 illustrates the steps involved in the error backpropropation training of multilayer NNs where the black parts show the flow of signals in feedforward (recall) mode and the gray parts depict the flow of errors in the training mode. In the *error backpropagation training procedure*, in the feedforward phase a single pattern x is submitted to the network to generate the layers response y. In training mode the error

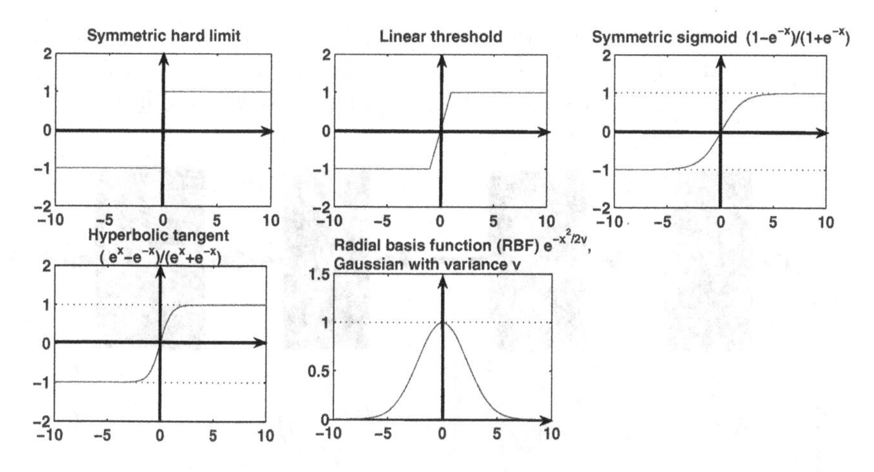

Fig. 1.2 Some common choices for the activation functions of multilayer neural networks

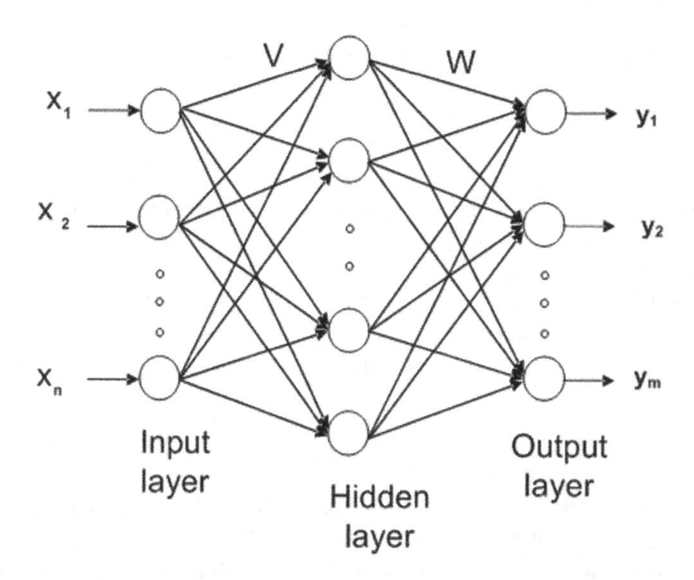

Fig. 1.3 A three layer neural network.

of the output layer, i.e., $\|y_d - y\|$ is propagated towards the network input layer and the weights are adjusted based on the steepest decent gradient method accordingly. The updating rules for the output and the hidden layer weights specified in equation

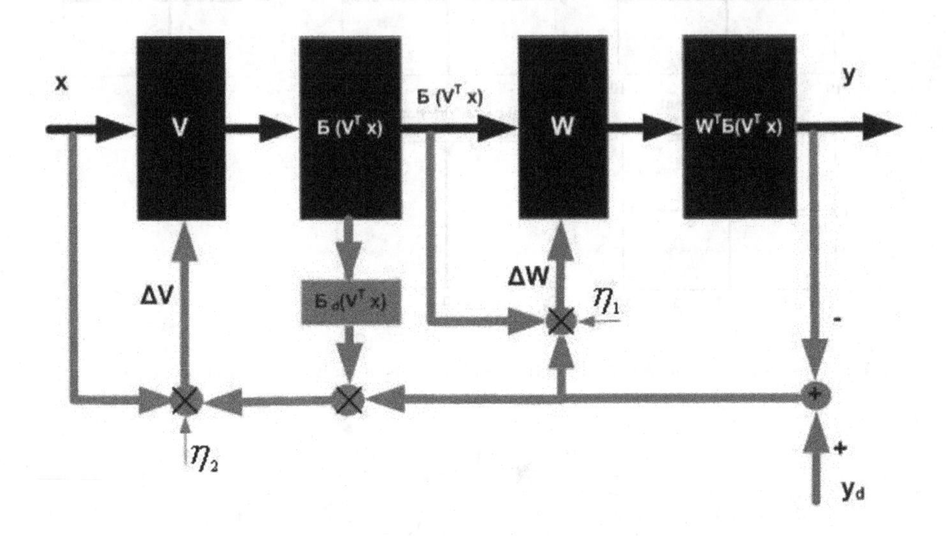

Fig. 1.4 A three layer neural network (two weighting layer) training by error BP algorithm [42].

(1.16) are defined as follows:

$$\dot{W} = -\eta_1(\frac{\partial \bar{E}}{\partial W}) \tag{1.17}$$

$$\dot{V} = -\eta_2(\frac{\partial \bar{E}}{\partial V}), \tag{1.18}$$

where η_1 and η_2 are the learning rates that are selected by the designer, $\bar{E} = \frac{1}{2}E^T E$ and $E = y_d - y$ is the output error. By considering that the activation function of the output layer is linear and after performing some algebraic manipulations, the updating rules can be simplified to

$$\dot{W} = -\eta_1 \sigma(V^T x) E^T \tag{1.19}$$

$$\dot{V} = -\eta_2 x(\sigma_d W E)^T, \tag{1.20}$$

where σ_d is the derivative of the hidden layer activation function, $\sigma(.)$. For more details regarding the backpropagation algorithm refer to [42].

A multilayer NNs can be employed for estimating the states of nonlinear systems having uncertain and unmodelled dynamics. The following section presents results on the neuro-observer specifications.

1.2.1.1 Neural Network Observer Classifications

Unlike most classic observers, neuro-observers estimate the states of the system despite the fact that the nonlinear dynamics of the system is also assumed to be unknown *a priori*. In fact, a neural network observer can be considered as a combination of a NN and a classic observer. The performance of these observers can be evaluated based on several criterion as listed below:

- **Type of the classic observer:** The performance of a neuro-observer can be investigated based on the type and the structure of the employed classic observer. For instance a neuro-observer can be considered as combination of a neural network to estimate the unknown nonlinear function of the system dynamics and a Luneberger observer, or a Kalman filter which its gains are estimated by a neural network.
- **The structure of neural network:** In view of Theorem 1.1, by using a multi-layer NN for some sufficiently large number of hidden layer neurons, there exist weights and thresholds so that a continuous function $G(x)$ can be represented as

$$G(x) = W\sigma(Vx+b) + \varepsilon(x), \tag{1.21}$$

where W and V are the weight matrices of the output and the hidden layers, respectively, $\sigma(.)$ is the hidden layer activation function, b is the threshold, and $\varepsilon(x)$ is the bounded neural network approximation error. The so-called Linear-in-Parameters Neural Network (LPNN) is obtained by fixing the weights of the first layer, e.g. $V = I$. Hence, the function $G(x)$ can be expressed as

$$G(x) = W\sigma(x+b) + \varepsilon(x). \tag{1.22}$$

On the othe hand, when the neural network first layer weights are not fixed, i.e., as in equation (1.21) the resulting network is called Nonlinear-in-Parameter Neural Network (NLPNN).
It should be noted that eventhough analyzing LPNN is simpler when compared to NLPNN due to its *linear* parametrization, it cannot represent a large class of nonlinear systems. Therefore, considering the structure of a NN as linear or nonlinear-in-parameter is an important criterion in performance evaluation of a neuro-observer.
- **Stability of the neuro-observer:** Stability analysis of neuro-observers is usually performed by the Lyapunov's direct method. To design a stable neuro-observer two approaches are proposed in the literature:

  1. Designing a neuro-observer by employing the well-known and popular learning rule such as the backpropagation algorithm and then proposing a proper Lyapunov function candidate to guarantee the stability of the system [44].
  2. Defining a Lyapunov function candidate based on some (usually quadratic) functions of the weights and the estimation errors and then looking for a proper updating rule which can guarantee the negative definiteness of the

Lyapunov function time derivative and consequently stability of the system [45, 46].

Several neural network-based observers have been introduced in the literature. For instance, in [47] a general Multi-input Multi-output (MIMO) nonlinear system was linearized and an extended Kalman filter was used to estimate the states of the system. The gain of the proposed observer was computed by a multilayer feedforward neural network. A NLPNN observer was also proposed with a back stepping control for a class of nonlinear dynamics in [45]. In [46], a robust adaptive radial basis function neural network observer was introduced for a class of uncertain nonlinear systems in presence of time-varying unknown parameter with absolutely integrable and non-vanishing disturbance. By employing strictly positive real (SPR) systems and Lyapunov theory, a stable observer-based adaptive-neural control scheme was presented in [48] for non-affine nonlinear systems in the presence of unknown structure of nonlinearities.

In [36], neural network and sliding mode were utilized for designing a two-stage neural observer for a second-order mechanical system. In the first stage a dead-zone neural observer guarantees the boundedness of the observer error and then sliding mode technique is employed to assure convergence of the observer. In [49], a scheme using two separate *linear-in-parameter* (LPNN) neural networks was proposed to estimate the states of an *affine SISO nonlinear* system. Strong assumptions such as strictly positive real (SPR) condition were imposed on the output error equation. In addition, they considered *scalar* valued nonlinear functions in the system dynamics. This implies that the same nonlinear terms correspond to all the states. The observer presented in [50] is similar to the work introduced in [49] except that they have used only one neural network which is also linear-in-parameter for identification of all unknown nonlinear functions.

In [51], it was claimed that every general nonlinear model can be described by an affine model plus a bounded unmodeled dynamic term. Therefore, the affine model was used for observer design. However, no clear strategy was suggested for reducing the resulting modelling of errors, arbitrarily. In [52, 53], an observer for a general MIMO nonlinear system using a *linear-in-parameter* neural network was proposed and the SPR assumption was also relaxed. According to the authors, however, choosing proper values of the design parameters such as the various gains and functional links of the neural networks are extremely difficult. Moreover, the observer has an open-loop structure. Unlike the above mentioned observers, in [54] the weight-updating mechanism was based on the steepest descent gradient method. The observer was based on a general model of a MIMO nonlinear system and was shown to be experimentally stable, but no mathematical proof was given to support the experiments.

In the reminder of this section, the outline of the monograph is given.

1.3 Outline

1.3.0.2 Chapter 2: Neural Network-Based State Estimation Schemes

In this chapter, a stable neural network-based observer for general multivariable nonlinear systems is presented. In the first step, a linear-in-parameter neural network is used to approximate the nonlinear dynamics. The proposed neural network observer is extended to a nonlinear-in-parameter neural network structure which can be applied to systems with higher degrees of nonlinearity without any *a priori* knowledge about the system dynamics. The learning rule for neural network is a novel approach based on the modified backpropagation algorithm. An e-modification term is added to guarantee robustness of the observer. No strictly positive real (SPR) or any other strong assumption is imposed for the proposed approach. The stability of the recurrent neural network observer is shown by Lyapunov's direct method. Simulation results for a flexible-joint manipulator are presented to demonstrate the enhanced performance capabilities that can be achieved by utilizing our proposed neural network observer.

1.3.0.3 CHAPTER 3: Neural Network-Based System Identification Schemes

The linear-in-parameter and nonlinear-in-parameter neural networks introduced in the previous chapter is applied in this chapter to design a stable identifier. By using Lyapunov's direct method, a novel and a more efficient approach is introduced to demonstrate the stability of the overall system. To show the performance of the proposed algorithm, an experimental set-up consisting of a three-link macro-micro manipulator is considered. The proposed approach is applied to identify the dynamics of the experimental robot. Experimental results are provided to show the effectiveness of the proposed learning scheme.

1.3.0.4 CHAPTER 4: An Actuator Fault Detection and Isolation: Experiments in Robotic Manipulators

This chapter presents an actuator fault detection and identification (FDI) scheme for nonlinear systems. A state space approach is used in which the states are assumed to be available. Moreover, the actuators are assumed to be faulty with bias faults. The (NLPNN) identifier introduced in Chapter 3 is employed to identify the unknown fault. The FDI scheme is based on a hybrid model (composed of an analytical nominal model and a neural network model) of the nonlinear system. To evaluate the performance of the proposed fault detection scheme, two case studies are considered. First, FDI in a spacecraft attitude control system with reaction wheel type of actuators is demonstrated via simulations. Next, the FDI problem in robotic manipulators is considered. Experimental results are presented to demonstrate the effectiveness of the proposed fault detection scheme.

1.3.0.5 CHAPTER 5: A Robust Actuator Gain Fault Detection and Isolation Scheme

The problem of actuator gain fault detection and isolation scheme for nonlinear systems is addressed in this chapter. The proposed FDI approach employs a neural network-based observer to detect and identify the severity of *actuator gain* faults in the presence of disturbances and uncertainties in model and sensory measurements. The neural network weights are updated based on a modified *dynamic* backpropagation scheme. The proposed FDI scheme does not rely on the availability of all the state measurements. In a number of methods in the literature, the fault function acts as an additive (bias) term on the actuator, whereas in the proposed FDI scheme the fault function acts as a multiplicative (gain) term which makes the stability and convergence of the overall FDI scheme rather challenging. The stability of the overall fault detection approach in the presence of unknown actuator gain fault as well as plant and sensor uncertainties is shown using Lyapunov's direct method with no restrictive assumptions on the system and/or the FDI algorithm. The performance of the proposed FDI approach is evaluated via simulations performed on reaction wheel type actuators that are commonly utilized in the attitude control subsystems (ACS) of satellites.

1.3.0.6 CHAPTER 6: A Robust Sensor and Actuator Fault Detection and Estimation

This chapter addresses a robust actuator and sensor fault detection and isolation scheme for a general nonlinear system using a neural network-based observer. The nonlinear system is subject to state and sensor uncertainties and disturbances. Two recurrent neural networks are employed to identify general unknown actuator and sensor faults. The neural network weights are updated based on a modified backpropagation scheme. Unlike many previous methods presented in the literature, the proposed FDI scheme does not rely on the availability of all the state measurements. It should be mentioned that since the two neural networks are coupled, studying the stability of the overall system is very challenging. The stability of the overall fault detection and isolation approach in the presence of unknown sensor and actuator faults as well as plant and sensor noise and uncertainties is shown by using Lyapunov's direct method. Stability analysis developed in this chapter imposes no restrictive assumptions on the system and/or the FDI algorithm. Magnetorquer type actuators and Magnetometer type sensors that are commonly utilized in the attitude determination and control of Low-Earth Orbit (LEO) satellites are considered for case studies. The effectiveness of the proposed fault diagnosis strategy is demonstrated through extensive simulation studies.

Chapter 2
Neural Network-Based State Estimation Schemes

2.1 Introduction

In this chapter, two neural network-based adaptive observers for a general model of MIMO nonlinear systems are proposed. The first proposed neural network is linear-in-parameter and the second one is *nonlinear* in its parameters which makes it applicable to many systems with arbitrary degrees of nonlinearity and complexity. The online weight-updating mechanism is a modified version of the backpropagation algorithm with a simple structure together with an e-modification term added to guarantee the robustness of the observer. The stability of the recurrent neural network observers are shown by Lyapunov's direct method. Moreover, the strictly positive real (SPR) assumption imposed on the output error equation is relaxed.

The reminder of this chapter is organized as follows: in Section 2.2, the observer problem is stated and the general structure of the neuro-observer is given. The proposed *linear-in-parameter* neuro-observer is introduced in Section 2.3. In Section 2.4, the results are extended to the case of *nonlinear-in-parameter* neuro-observer. Section 2.5 gives a model of flexible-joint manipulators. The observer performance is evaluated and demonstrated in Section 2.6 by simulation carried out on single- and two-link flexible-joint manipulators. Section 2.7 provides brief conclusions of this chapter.

2.2 Problem Formulation

Consider the general model of a nonlinear MIMO system

$$\begin{aligned} \dot{x}(t) &= f(x,u) \\ y(t) &= Cx(t), \end{aligned} \tag{2.1}$$

H.A. Talebi et al., *Neural Network-Based State Estimation of Nonlinear Systems*,
Lecture Notes in Control and Information Sciences 395,
DOI 10.1007/978-1-4419-1438-5_2,

where $u \in R^m$ is the input, $y \in R^m$ is the output, $x \in R^n$ is the state vector of the system, and $f(.,.)$ is a vector-valued unknown nonlinear function.

The objective is to estimate the state x in the presence of unknown function $f(x,u)$ in the system dynamics (2.1). The following assumptions will help us in deriving proper update rules as well as in stability analysis:

Assumption 2.1 *The nonlinear system (2.1) is observable.*

Assumption 2.2 *Another reasonable assumption made here is that the open-loop system is stable. In other words, the states of the system, $x(t)$ are bounded in L_∞ which is a common assumption in identification schemes.*

Now, selecting a Hurwitz matrix, A such that the pair (C,A) is observable and adding Ax to and subtracting it from (2.1) yields

$$\begin{aligned} \dot{x}(t) &= Ax + g(x,u) \\ y(t) &= Cx(t), \end{aligned} \tag{2.2}$$

where $g(x,u) = f(x,u) - Ax$.

The key to designing a neuro-observer is to employ a neural network to identify the nonlinearity and a conventional observer to estimate the states. By invoking a Luenberger observer [21], the observer model of the system (2.2) can be defined as follows

$$\begin{aligned} \dot{\hat{x}}(t) &= A\hat{x} + \hat{g}(\hat{x},u) + G(y - C\hat{x}) \\ \hat{y}(t) &= C\hat{x}(t), \end{aligned} \tag{2.3}$$

where $\hat{x}$ denotes the state of the observer, and the observer gain $G \in R^{n\times m}$ is selected such that $A - GC$ becomes a Hurwitz matrix. It should be noted that the gain G is guaranteed to exist, since A can be selected such that (C,A) is observable.

The structure of a neuro-observer is shown in Fig. (2.1). In this figure, $\hat{x}$ denotes the state of the recurrent model (2.3). Corresponding to the Hurwitz matrix A, $M(s) := (sI - A)^{-1}$ is also shown which is an $n \times n$ matrix whose elements are stable transfer functions. To approximate the nonlinear function $g(x,u)$ a multilayer NN is considered. According to Theorem 1.1, a multilayer NN with sufficiently large number of hidden layer neurons can estimate the unknown function $g(x,u)$ as follows:

$$g(x,u) = W^T\sigma(V^T\bar{x}^T) + \varepsilon(x),$$

where W and V are the weight matrices of the output and hidden layers, respectively, $\bar{x} = [x \quad u]$, $\varepsilon(x)$ is the bounded neural network approximation error, and $\sigma(.)$ is the transfer function of the hidden neurons that is usually considered as a tangent hyperbolic function presenting below:

$$\sigma_i(V_i\bar{x}) = \frac{2}{1 + exp^{-2V_i\bar{x}}} - 1, \tag{2.4}$$

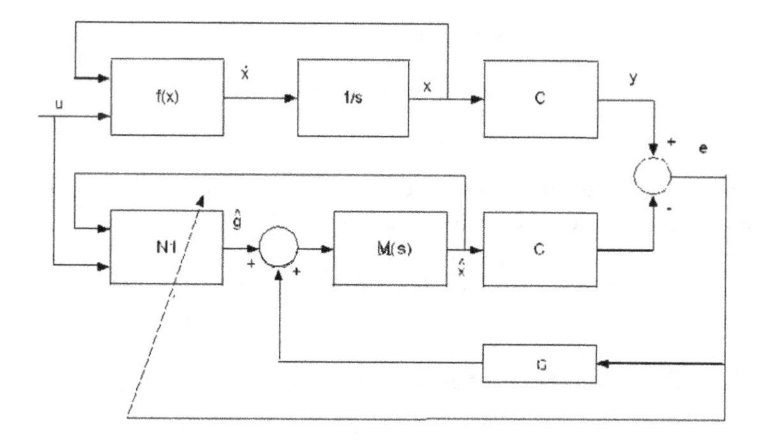

Fig. 2.1 The structure of the proposed neural network observer.

where V_i is the ith row of V, and $\sigma_i(V_i\bar{x})$ is the ith element of $\sigma(V\bar{x})$.

It is assumed that the upper bound on fixed ideal weights W and V exist such that

$$\|W\|_F \leq W_M \tag{2.5}$$

$$\|V\|_F \leq V_M. \tag{2.6}$$

It is also known that the tangent hyperbolic function is bounded by σ_m, i.e.,

$$\|\sigma(V\bar{x})\| \leq \sigma_m. \tag{2.7}$$

In the following two sections, a LPNN and a NLPNN observer are introduced and the stability of the observers are studied by Lyapunov's direct method.

2.3 Linear-in-Parameter Neural Network-Based Observer

In this section a stable LPNN neuro-observer is proposed for nonlinear systems whose dynamics are governed by (2.2). Recalling from Section 1.2.1 that, the linear-in-parameter neural network (LPNN) is obtained by fixing the weights of the first layer as $V = I$. Then, the model can be expressed as

$$g(x,u) = W^T\sigma(\bar{x}) + \varepsilon. \tag{2.8}$$

Thus, the function g can be approximated by an LPNN as

$$\hat{g}(\hat{x},u) = \hat{W}^T\sigma(\hat{\bar{x}}). \tag{2.9}$$

The proposed observer is then given by

$$\begin{aligned}\dot{\hat{x}}(t) &= A\hat{x} + \hat{W}^T\sigma(\hat{\bar{x}}) + G(y - C\hat{x})\\ \hat{y}(t) &= C\hat{x}(t).\end{aligned} \tag{2.10}$$

On the other hand, by defining the state estimation error as $\tilde{x} = x - \hat{x}$ and using (2.2), (2.9) and (2.10), the error dynamics can be expressed as

$$\begin{aligned}\dot{\tilde{x}}(t) &= Ax + W\sigma(\bar{x}) - A\hat{x} - \hat{W}\sigma(\hat{\bar{x}}) - G(Cx - C\hat{x}) + \varepsilon(x)\\ \tilde{y}(t) &= C\tilde{x}(t).\end{aligned} \tag{2.11}$$

Now, adding $W\sigma(\hat{\bar{x}})$ to and subtracting from (2.11) leads to

$$\begin{aligned}\dot{\tilde{x}}(t) &= A_c\tilde{x} + \tilde{W}\sigma(\hat{\bar{x}}) + w(t)\\ \tilde{y}(t) &= C\tilde{x}(t),\end{aligned} \tag{2.12}$$

where $\tilde{W} = W - \hat{W}$, $A_c = A - GC$, and $w(t) = W[\sigma(\bar{x}) - \sigma(\hat{\bar{x}})] + \varepsilon(x)$ is a bounded disturbance term i.e., $\|w(t)\| \leq \bar{w}$ for some positive constant $\bar{w}$, due to the tangent hyperbolic function and the boundedness of the ideal neural network weights, W.

Once the structure of the neural network is known, a proper learning rule should be defined to train the network. This weight-updating mechanism is usually defined in such a way that the stability of the observer is guaranteed. Furthermore, the adaptive law should not be complicated or limited by some strong constraints. Backpropagation (BP) is one of the most popular algorithms that has been widely used for classification, recognition, identification, observation, and control problems. BP owes its popularity to the simplicity in structure which makes it a viable choice for practical problems. However, the main drawback of the previous work (e.g. [55] and [56]) is the lack of a mathematical proof of stability. The following theorem provides a stable neuro-observer by proposing an weight-updating mechanism based on the modified backpropagation algorithm plus an e-modification term to guarantee its robustness.

Theorem 2.1. *Consider the plant model (2.1) and the observer model (2.10). Given Assumptions 2.1 and 2.2, if the weights of the LPNN are updated according to*

$$\dot{\hat{W}} = -\eta(\tilde{y}^T CA_c^{-1})^T(\sigma(\hat{\bar{x}}))^T - \rho\|\tilde{y}\|\hat{W}, \tag{2.13}$$

where $\eta > 0$ is the learning rate, $J = \frac{1}{2}(\tilde{y}^T\tilde{y})$ is the objective function and ρ is a small positive number, then $\tilde{x}, \tilde{W}, \tilde{y} \in L_\infty$

Proof: To prove this theorem, it is firstly shown that the first term in (2.13) is the backpropagation term and the second term is the e-modification terms for incorporating damping in the equations, i.e.,

$$\dot{\hat{W}} = -\eta(\frac{\partial J}{\partial \hat{W}}) - \rho\|\tilde{y}\|\hat{W}. \tag{2.14}$$

By employing the chain rule $\frac{\partial J}{\partial \hat{W}}$ can be computed as follows

$$\frac{\partial J}{\partial \hat{W}} = \frac{\partial J}{\partial \tilde{y}} \frac{\partial \tilde{y}}{\partial \hat{x}} \frac{\partial \hat{x}}{\partial \hat{W}} = -\tilde{y}^T C \frac{\partial \hat{x}}{\partial \hat{W}}. \tag{2.15}$$

The above equation represent a set of nonlinear dynamical system and the so-called backpropagation in time (dynamic backpropagation) should be utilized for solving the gradient $\frac{\partial \hat{x}}{\partial \hat{W}}$. However, this adds to the complexity of the observer and makes the real-time implementation of the approach very difficult. To cope with this problem, it is suggested to use the static approximation of the gradient, i.e., by setting $\dot{\hat{x}} = 0$ in (2.10), one achieves

$$\frac{\partial \hat{x}}{\partial \hat{W}} \approx -A_c^{-1} \frac{\partial g}{\partial \hat{W}}. \tag{2.16}$$

Now by using (2.9), (2.15) and (2.16), the learning rule (2.14) can be written as

$$\dot{\hat{W}} = -\eta (\tilde{x}^T C^T C A_c^{-1})^T (\sigma(\hat{x}))^T - \rho \|C\tilde{x}\| \hat{W}. \tag{2.17}$$

Therefore, the dynamics of the weight error $\tilde{W} = W - \hat{W}$ can be expressed as

$$\dot{\tilde{W}} = \eta (\tilde{x}^T C^T C A_c^{-1})^T (\sigma(\hat{x}))^T \rho \|C\tilde{x}\| \hat{W}. \tag{2.18}$$

To study the stability of the proposed observer let us consider the Lyapunov function candidate as below

$$L = \frac{1}{2} \tilde{x}^T P \tilde{x} + \frac{1}{2} tr(\tilde{W}^T \rho^{-1} \tilde{W}), \tag{2.19}$$

where $P = P^T$ is a positive-definite matrix satisfying

$$A_c^T P + P A_c = -Q, \tag{2.20}$$

for the Hurwitz matrix A_c and some positive-definite matrix Q. Then, taking time derivative of (2.19) along the trajectories (2.12) yields

$$\dot{L} = \frac{1}{2} \dot{\tilde{x}}^T P \tilde{x} + \frac{1}{2} \tilde{x}^T P \dot{\tilde{x}} + tr(\tilde{W}^T \rho^{-1} \dot{\tilde{W}}). \tag{2.21}$$

Now, by substituting (2.12),(2.20), and (2.18) into (2.21), one can get

$$\begin{aligned} \dot{L} &= -\frac{1}{2} \dot{\tilde{x}}^T Q \tilde{x} + \tilde{x}^T P (\tilde{W} \sigma(\hat{x}) + w) \\ &+ tr(-\tilde{W}^T l \tilde{x} \sigma^T + \tilde{W}^T \|C\tilde{x}\| (W - \tilde{W})), \end{aligned} \tag{2.22}$$

where $l = \eta \rho^{-1} A_c^{-T} C^T C$. On the other hand, we have

$$tr(\tilde{W}^T(W-\tilde{W}) \leq W_M\|\tilde{W}\| - \|\tilde{W}\|^2 \quad (2.23)$$
$$tr(\tilde{W}^T l\tilde{x}\sigma^T) \leq \sigma_m\|\tilde{W}^T\| \; \|l\| \; \|\tilde{x}\|, \quad (2.24)$$

where W_M and σ_m are given by (2.5) and (2.7). Note that the last inequality in (2.23) is obtained by using the fact that for two column vectors A and B, the following equality holds:

$$tr(AB^T) = B^T A. \quad (2.25)$$

Now, by using (2.23) and (2.24), one can get

$$\dot{L} \leq -\frac{1}{2}\lambda_{min}(Q)\|\tilde{x}\|^2 + \|\tilde{x}\|\|P\|(\|\tilde{W}\|\sigma_m + \bar{w})$$
$$+ \sigma_m\|\tilde{W}\|\|l\|\|\tilde{x}\| + (W_M\|\tilde{W}\| - \|\tilde{W}\|^2)\|C\|\|\tilde{x}\|.$$

Furthermore, by completing the squares of $\|\tilde{W}\|$ the following condition is obtained to ensure negative definiteness of $\dot{L}$:

$$\|\tilde{x}\| \geq (2\|P\|\bar{w} + (\sigma_m\|p\| + W_M\|C\| + \sigma_m\|l\|)^2/2)/\lambda_{min}(Q).$$

In fact, $\dot{L}$ is negative definite outside the ball with radius b described as $\chi = \{\tilde{x} \mid \|\tilde{x}\| > b\}$, and $\tilde{x}$ is uniformly ultimately bounded. The region inside the ball is attractive, since the increase of $\dot{L}$ for smaller values of $\|\tilde{x}\|$ will increase L and $\tilde{x}$, which brings the $\tilde{x}$ outside the ball χ where $\dot{L}$ is negative semi-definite and results in reducing L and $\tilde{x}$. The above analysis shows the ultimate boundedness of $\tilde{x}$. Further details concerning the notion of ultimate boundedness is given in Appendix A.

To show the boundedness of the weight error $\tilde{W}$, consider (2.18) which can be rewritten as

$$\dot{\tilde{W}} = f_1(\tilde{x}) + \rho\|C\tilde{x}\|\hat{W} = f_1(\tilde{x}) + \alpha_1 W - \alpha_1\tilde{W}, \quad (2.26)$$

where

$$f_1(\tilde{x}) = \eta(\tilde{x}^T C^T C A_c^{-1})^T(\sigma(\hat{x}))^T$$
$$\alpha_1 = \rho\|C\tilde{x}\|.$$

It can be seen that $f_1(.)$ is bounded since $\tilde{x}$ and $\sigma(\hat{V}\hat{x})$ are both bounded, C is bounded, and A_c is a Hurwitz matrix . Given the fact the the ideal weight W is fixed, (2.26) can be regarded as a linear system with bounded input $(f_1(\tilde{x}) + \alpha_1 W)$. It is clear that this system is stable since α_1 is positive and the system input remains bounded. Hence, the boundedness of $\tilde{W}$ is also ensured. This completes the proof of the theorem. □

A stable LPNN neuro-observer was introduced in this section. The next section provides a neuro-observer for larger class of nonlinear dynamics, called nonlinear-in-parameter (NLPNN) neuro-observer.

2.4 Nonlinear-in-Parameter Neural Network-Based Observer

In more general neuro-observer structure, where the weight matrix of the first layer is not restricted to be constant, the nonlinear function g can be approximated by a multilayer NN according to the following model:

$$\hat{g}(\hat{x},u) = \hat{W}\sigma(\hat{V}\hat{\bar{x}}). \tag{2.27}$$

Therefore, the observer dynamics can be given as

$$\begin{aligned}\dot{\hat{x}}(t) &= A\hat{x} + \hat{W}\sigma(\hat{V}\hat{\bar{x}}) + G(y - C\hat{x})\\ \hat{y}(t) &= C\hat{x}(t),\end{aligned} \tag{2.28}$$

and also the error dynamics can be expressed as

$$\begin{aligned}\dot{\tilde{x}}(t) &= A_c\tilde{x} + \tilde{W}\sigma(\hat{V}\hat{\bar{x}}) + w(t)\\ \tilde{y}(t) &= C\tilde{x}(t),\end{aligned} \tag{2.29}$$

where $\tilde{W} = W - \hat{W}$, $A_c = A - GC$, $w(t) = W[\sigma(V\bar{x}) - \sigma(\hat{V}\hat{\bar{x}})] + \varepsilon(x)$ is a bounded disturbance term i.e., $\|w(t)\| \leq \bar{w}$ for some positive constant $\bar{w}$, due to the tangent hyperbolic function and the boundedness of the ideal neural network weights V and W.

In Theorem 2.2, a learning rule is introduced for neuro-observer (2.28) which guarantees the stability of the observer and the boundedness of the error estimation.

Theorem 2.2. *Consider the plant model (2.1) and the observer model (2.28). Given Assumptions 2.1 and 2.2, if the weights of the NLPNN are updated according to*

$$\dot{\hat{W}} = -\eta_1(\tilde{y}^T CA_c^{-1})^T(\sigma(\hat{V}\hat{\bar{x}}))^T - \rho_1\|\tilde{y}\|\hat{W} \tag{2.30}$$

$$\dot{\hat{V}} = -\eta_2(\tilde{y}^T CA_c^{-1}\hat{W}(I - \Lambda(\hat{V}\hat{\bar{x}})))^T sgn(\hat{\bar{x}})^T - \rho_2\|\tilde{y}\|\hat{V}, \tag{2.31}$$

where $\Lambda(\hat{V}\hat{\bar{x}}) = diag\{\sigma_i^2(\hat{V}_i\hat{\bar{x}})\}, i = 1,2,...,m$ *and* $sgn(\hat{\bar{x}})$ *is the sign function:*

$$sgn(\hat{\bar{x}}) = \begin{cases} 1 & for\ \hat{\bar{x}} > 0\\ 0 & for\ \hat{\bar{x}} = 0\\ -1 & for\ \hat{\bar{x}} < 0\end{cases},$$

then $\tilde{x}, \tilde{W}, \tilde{V}, \tilde{y} \in L_\infty$, *i.e., the estimation error, weights error, and the output error are all bounded. In these equations,* $\eta_1, \eta_2 > 0$ *are the learning rates,* $J = \frac{1}{2}(\tilde{y}^T\tilde{y})$ *is the objective function and* ρ_1, ρ_2 *are positive numbers.*

Proof: Similar to the proof of Theorem 2.1, at first it is shown that the first terms in (2.30) and (2.31) are the backpropagation terms and the second terms are the e-modification which incorporates proper damping in the equations. Then, the stability of the proposed observer is studied by Lyapunov's direct method. In other words, the weights are updated based on the following dynamics

$$\dot{\hat{W}} = -\eta_1(\frac{\partial J}{\partial \hat{W}}) - \rho_1\|\tilde{y}\|\hat{W} \tag{2.32}$$

$$\dot{\hat{V}} = -\eta_2(\frac{\partial J}{\partial \hat{V}}) - \rho_2\|\tilde{y}\|\hat{V}. \tag{2.33}$$

Let us define

$$net_{\hat{v}} = \hat{V}\hat{\bar{x}} \tag{2.34}$$

$$net_{\hat{w}} = \hat{W}\sigma(\hat{V}\hat{\bar{x}}). \tag{2.35}$$

Therefore, $\frac{\partial J}{\partial \hat{W}}$ and $\frac{\partial J}{\partial \hat{V}}$ can be computed as [42]

$$\frac{\partial J}{\partial \hat{W}} = \frac{\partial J}{\partial net_{\hat{w}}}.\frac{\partial net_{\hat{w}}}{\partial \hat{W}}$$
$$\frac{\partial J}{\partial \hat{V}} = \frac{\partial J}{\partial net_{\hat{v}}}.\frac{\partial net_{\hat{v}}}{\partial \hat{V}}.$$

On the other hand, we have

$$\frac{\partial J}{\partial net_{\hat{w}}} = \frac{\partial J}{\partial \tilde{y}}.\frac{\partial \tilde{y}}{\partial \hat{x}}.\frac{\partial \hat{x}}{\partial net_{\hat{w}}} = -\tilde{y}^T C.\frac{\partial \hat{x}}{\partial net_{\hat{w}}} = -\tilde{x}^T C^T C.\frac{\partial \hat{x}}{\partial net_{\hat{w}}}$$
$$\frac{\partial J}{\partial net_{\hat{v}}} = \frac{\partial J}{\partial \tilde{y}}.\frac{\partial \tilde{y}}{\partial \hat{x}}.\frac{\partial \hat{x}}{\partial net_{\hat{v}}} = -\tilde{y}^T C.\frac{\partial \hat{x}}{\partial net_{\hat{v}}} = -\tilde{x}^T C^T C.\frac{\partial \hat{x}}{\partial net_{\hat{v}}}, \tag{2.36}$$

and

$$\frac{\partial net_{\hat{w}}}{\partial \hat{W}} = \sigma(\hat{V}\hat{\bar{x}})$$
$$\frac{\partial net_{\hat{v}}}{\partial \hat{V}} = \hat{\bar{x}}. \tag{2.37}$$

Now, by employing (2.28) and the definitions of $net_{\hat{v}}$ and $net_{\hat{w}}$ as defined in (2.34) and (2.35), respectively one can obtain

$$\frac{\partial \dot{\hat{x}}(t)}{\partial net_{\hat{w}}} = A_c\frac{\partial \hat{x}}{\partial net_{\hat{w}}} + I$$
$$\frac{\partial \dot{\hat{x}}(t)}{\partial net_{\hat{v}}} = A_c\frac{\partial \hat{x}}{\partial net_{\hat{v}}} + \hat{W}(I - \Lambda(\hat{V}\hat{\bar{x}})). \tag{2.38}$$

Similar to the approach adopted in the proof of Theorem 2.1, using the static approximation, yields in

$$\frac{\partial \hat{x}}{\partial net_{\hat{w}}} \approx -A_c^{-1}$$
$$\frac{\partial \hat{x}}{\partial net_{\hat{v}}} \approx -A_c^{-1}\hat{W}(I - \Lambda(\hat{V}\hat{\bar{x}})). \tag{2.39}$$

Then, substituting (2.36), (2.37), and (2.39), in the the learning rules (2.32) and (2.33) leads to

$$\dot{\hat{W}} = -\eta_1(\tilde{x}^T C^T C A_c^{-1})^T (\sigma(\hat{V}\hat{\bar{x}}))^T - \rho_1 \|C\tilde{x}\| \hat{W} \tag{2.40}$$

$$\dot{\hat{V}} = -\eta_2(\tilde{x}^T C^T C A_c^{-1} \hat{W}(I - \Lambda(\hat{V}\hat{\bar{x}})))^T \hat{\bar{x}}^T - \rho_2 \|C\tilde{x}\| \hat{V}. \tag{2.41}$$

Therefore, the learning rules (2.40) and (2.41) in terms of the weight errors $\tilde{W} = W - \hat{W}$ and $\tilde{V} = V - \hat{V}$, can be written as

$$\dot{\tilde{W}} = \eta_1(\tilde{x}^T C^T C A_c^{-1})^T (\sigma(\hat{V}\hat{\bar{x}}))^T + \rho_1 \|C\tilde{x}\| \hat{W} \tag{2.42}$$

$$\dot{\tilde{V}} = \eta_2(\tilde{x}^T C^T C A_c^{-1} \hat{W}(I - \Lambda(\hat{V}\hat{\bar{x}})))^T \hat{\bar{x}}^T + \rho_2 \|C\tilde{x}\| \hat{V}. \tag{2.43}$$

In order to simplify the stability analysis, we replace $\hat{\bar{x}}$ by $sgn(\hat{\bar{x}})$ in the above equation:

$$\dot{\tilde{V}} = \eta_2(\tilde{x}^T C^T C A_c^{-1} \hat{W}(I - \Lambda(\hat{V}\hat{\bar{x}})))^T sgn(\hat{\bar{x}})^T + \rho_2 \|C\tilde{x}\| \hat{V}. \tag{2.44}$$

As will be clear later, this modification is necessary to derive Equation (2.50), since $sgn(\hat{\bar{x}})$ is bounded but this is not necessarily true for $\hat{\bar{x}}$. Note that by using the sign of $\hat{\bar{x}}$, the weight update is guaranteed to move in the right direction. It can be seen that the learning rule (2.42) and (2.44) are equivalent to (2.30) and (2.31) expressed in terms of $\tilde{W}$ and $\tilde{V}$.

Now, to show ultimate boundedness of the estimation errors, let us consider the following positive definite Lyapunov function candidate

$$L = \frac{1}{2}\tilde{x}^T P \tilde{x} + \frac{1}{2} tr(\tilde{W}^T \tilde{W}) + \frac{1}{2} tr(\tilde{V}^T \tilde{V}), \tag{2.45}$$

where $P = P^T$ is a positive-definite matrix satisfying

$$A_c^T P + P A_c = -Q, \tag{2.46}$$

for the Hurwitz matrix A_c and some positive-definite matrix Q. The time derivative of (2.45) is given by

$$\dot{L} = \frac{1}{2}\dot{\tilde{x}}^T P \tilde{x} + \frac{1}{2}\tilde{x}^T P \dot{\tilde{x}} + tr(\tilde{W}^T \dot{\tilde{W}}) + tr(\tilde{V}^T \dot{\tilde{V}}). \tag{2.47}$$

Then, substituting (2.29), (2.42), (2.44) and (2.46) into (2.47) yields

$$\begin{aligned}\dot{L} = &-\frac{1}{2}\tilde{x}^T Q \tilde{x} + \tilde{x}^T P(\tilde{W}\sigma(\hat{V}\hat{\bar{x}}) + w) + tr(\tilde{W}^T l_1 \tilde{x} \sigma(\hat{V}\hat{\bar{x}})^T + \tilde{W}^T \rho_1 \|C\tilde{x}\|(W - \tilde{W})) \\ &+ tr(\tilde{V}^T (I - \Lambda(\hat{V}\hat{\bar{x}}))^T \hat{W}^T l_2 \tilde{x} sgn(\hat{\bar{x}})^T + \tilde{V}^T \rho_2 \|C\tilde{x}\|(V - \tilde{V})),\end{aligned} \tag{2.48}$$

where $l_1 = \eta_1 A_c^{-T} C^T C, l_2 = \eta_2 A_c^{-T} C^T C$. On the other hand, the following inequalities are always true

$$
\begin{aligned}
tr(\tilde{W}^T(W-\tilde{W})) &\leq W_M\|\tilde{W}\| - \|\tilde{W}\|^2 \\
tr(\tilde{V}^T(V-\tilde{V})) &\leq V_M\|\tilde{V}\| - \|\tilde{V}\|^2 \\
tr(\tilde{W}^T l_1 \tilde{x}\sigma(\hat{V}\hat{\bar{x}})^T) &\leq \sigma_m\|\tilde{W}^T\| \, \|l_1\| \, \|\tilde{x}\|,
\end{aligned}
\tag{2.49}
$$

where W_M, V_M and σ_m are given by (2.5)-(2.7). Now, using the facts that $\|\hat{W}\| = \|W-\tilde{W}\| \leq W_M + \|\tilde{W}\|$, $1-\sigma_m^2 \leq 1$, and (2.25) leads to the following inequality

$$
tr(\tilde{V}^T(I-\Lambda(\hat{V}\hat{\bar{x}}))^T\hat{W}^T l_2\tilde{x}sgn(\bar{x})^T) \leq \|\tilde{V}\|(W_M+\|\tilde{W}\|)\|l_2\|\|\tilde{x}\|. \tag{2.50}
$$

Then, by employing (2.49) and (2.50), one can get

$$
\begin{aligned}
\dot{L} \leq &-\frac{1}{2}\lambda_{min}(Q)\|\tilde{x}\|^2 + \|\tilde{x}\|\|P\|(\|\tilde{W}\|\sigma_m + \bar{w}) + \sigma_m\|\tilde{W}\|\|l_1\|\|\tilde{x}\| \\
&+ (W_M\|\tilde{W}\| - \|\tilde{W}\|^2)\rho_1\|C\|\|\tilde{x}\| + \|\tilde{V}\|\|l_2\|(W_M+\|\tilde{W}\|)\|\tilde{x}\| \\
&+ \rho_2\|C\|\|\tilde{x}\|(V_M\|\tilde{V}\| - \|\tilde{V}\|^2) = F.
\end{aligned}
\tag{2.51}
$$

By completing the squares for the terms involving $\|\tilde{W}\|$ and $\|\tilde{V}\|$, we look for conditions on $\|\tilde{x}\|$ which are independent of the neural network weights error and also make the time derivative of the Lyapunov function candidate, (2.45) negative. Toward this end, by defining $K_1 = \frac{\|l_2\|}{2}$ and adding $K_1^2\|\tilde{W}\|^2\|\tilde{x}\|$ and $\|\tilde{V}\|^2\|\tilde{x}\|$ to and subtracting them from the right hand side of (2.51), one can obtain

$$
\begin{aligned}
F = &-\frac{1}{2}\lambda_{min}(Q)\|\tilde{x}\|^2 + (\|P\|\bar{w} - (\rho_1\|C\| - K_1^2)\|\tilde{W}\|^2 - (K_1\|\tilde{W}\| - \|\tilde{V}\|)^2 \\
&+ (\|P\|\sigma_m + \sigma_m\,\|l_1\| + \rho_1\|C\|W_M)\|\tilde{W}\| + (\rho_2\|C\|V_M + \|l_2\|W_M)\|\tilde{V}\| \\
&- (\rho_2\|C\| - 1)\|\tilde{V}\|^2)\|\tilde{x}\|.
\end{aligned}
\tag{2.52}
$$

Next, let us introduce K_2 and K_3 as follows:

$$
\begin{aligned}
K_2 &= \frac{\rho_1 W_M\|C\| + \sigma_m\|l_1\| + \|P\|\sigma_m}{2(\rho_1\|C\| - K_1^2)} \\
K_3 &= \frac{\rho_2\|C\|V_M + \|l_2\|W_M}{2(\rho_2\|C\| - 1)}.
\end{aligned}
$$

Then, $K_2^2\|\tilde{x}\|$ and $K_3^2\|\tilde{x}\|$ are added to and subtracted from (2.52):

$$
\begin{aligned}
F = &-\frac{1}{2}\lambda_{min}(Q)\|\tilde{x}\|^2 + (\|P\|\bar{w} + (\rho_1\|C\| - K_1^2)K_2^2 + (\rho_2\|C\| - 1)K_3^2 \\
&- (\rho_1\|C\| - K_1^2)(K_2 - \|\tilde{W}\|)^2 - (\rho_2\|C\| - 1)(K_3 - \|\tilde{V}\|)^2 \\
&- (K_1\|\tilde{W}\| - \|\tilde{V}\|)^2)\|\tilde{x}\|.
\end{aligned}
$$

Now, assuming the $\rho_1 \geq \frac{K_1^2}{\|C\|}$, $\rho_2 \geq \frac{1}{\|C\|}$ and using the fact that the last three terms of (2.53) are negative yields

$$F \leq -\frac{1}{2}\lambda_{min}(Q)\|\tilde{x}\|^2 + \|\tilde{x}\|(\|P\|\bar{w} + (\rho_1\|C\| - K_1^2)K_2^2 + (\rho_2\|C\| - 1)K_3^2). \quad (2.53)$$

Therefore, the following condition on $\|\tilde{x}\|$ guarantees the negative semi-definiteness of $\dot{L}$:

$$\|\tilde{x}\| > \frac{2(\|P\|\bar{w} + (\rho_1\|C\| - K_1^2)K_2^2 + (\rho_2\|C\| - 1)K_3^2)}{\lambda_{min}(Q)} = b. \quad (2.54)$$

In fact, $\dot{L}$ is negative definite outside the ball with radius b described as $\chi = \{\tilde{x} \mid \|\tilde{x}\| > b\}$, and $\tilde{x}$ is uniformly ultimately bounded.

To study the boundedness of the weight errors $\tilde{W}$ and $\tilde{V}$, let us rewrite (2.42) and (2.44) as

$$\dot{\tilde{W}} = f_1(\tilde{x}, \hat{V}) + \rho_1\|C\tilde{x}\|\hat{W} = f_1(\tilde{x}, \hat{V}) + \alpha_1 W - \alpha_1 \tilde{W} \quad (2.55)$$

$$\dot{\tilde{V}} = f_2(\tilde{x}, \hat{W}, \hat{V}) + \rho_2\|C\tilde{x}\|\hat{V} = f_2(\tilde{x}, \hat{W}, \hat{V}) + \alpha_2 V - \alpha_2 \tilde{V}, \quad (2.56)$$

where

$$\begin{aligned} f_1(\tilde{x}, \hat{V}) &= \eta_1(\tilde{x}^T C^T C A_c^{-1})^T (\sigma(\hat{V}\hat{\bar{x}}))^T \\ f_2(\tilde{x}, \hat{W}, \hat{V}) &= \eta_2(\tilde{x}^T C^T C A_c^{-1}\hat{W}(I - \Lambda(\hat{V}\hat{\bar{x}})))^T sgn(\hat{\bar{x}})^T \\ \alpha_1 &= \rho_1\|C\tilde{x}\| \\ \alpha_2 &= \rho_2\|C\tilde{x}\|. \end{aligned}$$

Since $\tilde{x}$ and $\sigma(\hat{V}\hat{\bar{x}})$ are both bounded, C is bounded, and A_c is a Hurwitz matrix, it can be concluded that $f_1(.)$ is bounded. Given that the ideal weight W is fixed and α_1 is positive, (2.55) can be regarded as a linear system with bounded input $(f_1(\tilde{x}, \hat{V}) + \alpha_1 W)$ and therefore, it can be assured that $\tilde{W}$ is bounded. Given that $\tilde{W} \in L_\infty$, it can be observed that $f_2(.)$ is also bounded since all its arguments are bounded including $\Lambda(.)$ as defined below equation (2.31). Consequently, similar analysis shows that (2.56) also represents a stable bounded input linear system and hence $\tilde{V} \in L_\infty$. The key to the above analysis is that $\tilde{V}$ only appears in $f_1(.)$ and $f_2(.)$ as bounded functions ($\sigma(.)$ and $\Lambda(.)$). This completes the proof. □

It is worth mentioning that the size of the estimation error bound b can be kept small by proper selection of the damping factors, A_c and the learning rates (through K_2 and K_3) such that a higher accuracy can be achieved. It should be noted that since ρ_1 and ρ_2 are design parameters, the conditions on them do not restrict the applicability of the proposed approach.

Remark 2.1 *In many cases, not all system states directly appear in the output of the system. Hence, some elements of C would be zero and this will slow down the learning process because of the structure of the backpropagation algorithm (see equations (2.42) and (2.44)). It is suggested that for the purpose of training only, the output matrix C is modified to C_1 such that all the states appear in the output of the system and directly contribute to the observation error, i.e., (2.30) and (2.31)* *can be redefined as*

$$\dot{\hat{W}} = -\eta_1(\tilde{y}^T C_1 A_c^{-1})^T (\sigma(\hat{V}\hat{\bar{x}}))^T - \rho_1 \|\tilde{y}\| \hat{W} \tag{2.57}$$

$$\dot{\hat{V}} = -\eta_2(\tilde{y}^T C_1 A_c^{-1} \hat{W}(I - \Lambda(\hat{V}\hat{\bar{x}})))^T sgn(\hat{\bar{x}})^T - \rho_2 \|\tilde{y}\| \hat{V}. \tag{2.58}$$

However, the rest of the proof and conditions remain unchanged.

Remark 2.2 *There are different parameters that can be adjusted to control the convergence of the neural network e.g. the learning rates η_1 and η_2, the damping factors ρ_1 and ρ_2, and the Hurwitz matrix A. It is well known that larger learning rates can lead to faster convergence but extra care should be taken to avoid overshoot. Moreover, although increasing the damping factor can improve the stability of the system, too much damping can lead to premature convergence of the weights which might be far from the ideal weights. The Hurwitz matrix A which is primarily used for stable integration has considerable effect on convergence as well as accuracy of the state estimation. One has to be concerned with different issues when selecting this matrix. First, it should be selected such that the pair (C,A) is observable. Since, the states of the estimator are obtained through $M(s)$ whose poles are the eigenvalues of A, a more stable matrix (matrix with eigenvalues farther to the left in the complex plane) helps the states of the estimator track those of the actual system with better accuracy (in terms of delays). However, a more stable A and hence A_c might slow down the convergence of the weights since A_c^{-1} is used in updating the weights (see equations (2.42) and (2.44)). One solution is to use a more stable A for better accuracy and use higher learning rates for better convergence.*

2.5 A Case Study: Application to State Estimation of Flexible-Joint Manipulators

The state estimation of flexible-joint manipulators is considered as a case study to evaluate the performance of the proposed observer. Robot manipulators with joint or link flexibility are proper examples of systems with high nonlinearity, unmodeled dynamics, and parameter variations.

A dominant source of compliance in robotic systems is the result of flexibility in the motor transmissions. . For instance, the unconventional gear-tooth meshing action of the harmonic drive makes it possible to acquire higher gear ratio and high torque capability in a compact geometry. On the other hand, a harmonic drive transmission is much more flexible than a conventional gear transmission. In Appendix B, the harmonic drives are explained in more details. The flexibility of the joint causes difficulty in modeling manipulator dynamics and becomes a potential source of uncertainty that can degrade the performance of a manipulator and in some cases can even destabilize the system [38]. Consequently, addressing this issue is important for calibration as well as modeling and control of robot manipulators. Joint

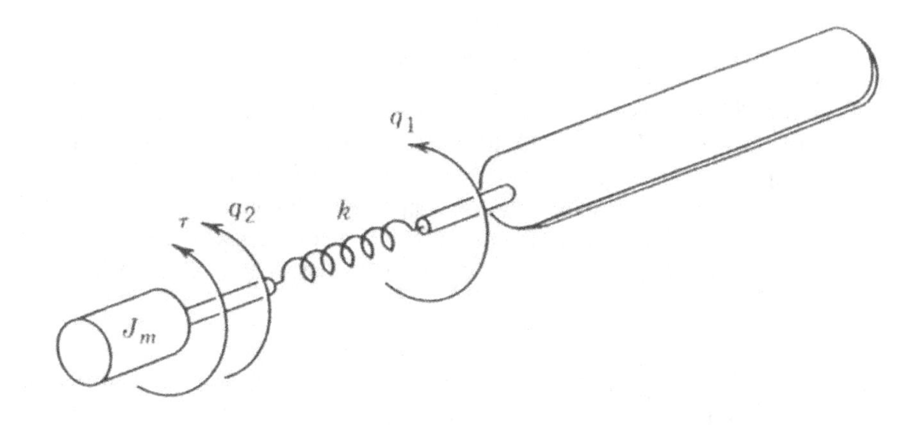

Fig. 2.2 The schematic of flexible-joint manipulator modeled by torsional spring [3]

elasticity can be modeled as a torsional spring between the input shaft (motor) and the output shaft (link) of the manipulator, as shown in Fig. 2.2. Due to the presence of joint flexibility, there are twice as many degrees of freedom compared to the rigid joint case.

To compensate for joint flexibility, many sophisticated control algorithms have been proposed both in constrained [57, 58] and unconstrained motions [59, 60, 61, 62, 63, 64, 65]. In [65], an adaptive control scheme is addressed for flexible-joint robots. Most of these schemes however, assume the availability of both the link and the motor positions, a condition that may not always be satisfied. Luenberger observers, reduced-order high-gain observers, and Kalman filter based observers have been used to relax the requirement of measurement from both sides of the transmission device [66, 67, 68, 28, 69, 70]. However, a fundamental assumption underlying all of these methods is that the system nonlinearities are completely known *a priori*. In this section, our proposed neural network observer is applied to a flexible-joint robot when the motor positions and velocities are available and the link positions and velocities need to be estimated. This choice of measured variables is the most practical.

Although, no *a priori* knowledge about the system dynamics is required for our state estimation approach, the analytical model of the manipulator is still needed for simulation purposes. In the following section, a dynamic model of a flexible-joint manipulator is introduced.

2.5.1 Manipulator Model

Using the Lagrangian approach, the flexible-joint manipulator can be modeled by the following equations [3].

$$\begin{aligned} D_l(q_1)\ddot{q}_1 + C_1(q_1,\dot{q}_1) + g(q_1) + B_1\dot{q}_1 &= \tau_s \\ J\ddot{q}_2 + \tau_s + B_2\dot{q}_2 &= \tau, \end{aligned} \tag{2.59}$$

where $q_1 \in \Re^n$ is the vector of link positions, $q_2 \in \Re^n$ is the vector of motor shaft positions, $g(q_1) \in \Re^n$ is the gravity loading force, $C_1(q_1,\dot{q}_1) \in \Re^n$ is the term corresponding to the centrifugal and Coriolis forces, $B_1 \in \Re^{n\times n}$ and $B_2 \in \Re^{n\times n}$ are the viscous damping matrices at the output and input shafts respectively, $D_l(q_1) \in \Re^{n\times n}$ and $J \in \Re^{n\times n}$ are the robot and the actuator inertia matrices respectively, and τ is the input torque. The reaction torque τ_s from the rotational spring is often considered as

$$\tau_s = K(q_2 - q_1) + \beta(q_1,\dot{q}_1,q_2,\dot{q}_2),$$

where $K \in \Re^{n\times n}$ is the positive-definite stiffness matrix of the rotational spring that represents the flexibility present between the input and the output shafts. In general, there is an unknown nonlinear force $\beta(q_1,\dot{q}_1,q_2,\dot{q}_2)$ which can be regarded as a combination of a nonlinear spring and friction at the output shafts of the manipulator. The reaction torque τ_s cannot be modeled accurately and is assumed to be unknown for observer design and is included for simulation purposes only. For more details regarding flexible-joint manipulator dynamics refer to Appendix B.

It should be noted that the flexible-joint manipulator system is marginally stable, i.e., its states are bounded for all bounded inputs except the step input. All results except those obtained for generalization are obtained for sinusoidal inputs. However, it is well known that the flexible-joint system can be stabilized using PD control of joint position feedback with arbitrary positive gains [3, 71]. Hence, a joint PD controller can be used to make the system stable without any *a priori* knowledge about the system dynamics. The joint position measurements are readily available. Hence, for a practical application this remedy can be used as in [72]. This technique is utilized for simulations concerning the generalization issue.

2.6 Simulation Results

In this section, the performance of the proposed observer is investigated on single-link and two-link flexible-joint manipulators.

2.6.1 A Single-link Flexible-Joint Manipulator

Consider a single-link flexible-joint manipulator, (2.59) whose its state vector is defined as $x = [q_1\ \dot{q}_1\ q_2\ \dot{q}_2]$ and numerical value of its parameters are given below:

$$\begin{aligned} &J = 1.16kg.m^2,\ m = 1Kg,\ l = 1m,\ K = 100N/m \\ &\eta = 10,\ \rho = 1.5, \end{aligned}$$

$C = \begin{bmatrix} 1\,0\,0\,0 \\ 0\,1\,0\,0 \end{bmatrix}$, and A is an 4×4 block diagonal matrix whose blocks A_{ii} are selected as $\begin{bmatrix} -20 & 1 \\ 0 & -20 \end{bmatrix}$, $i = 1...2$, where J is the motor inertia, m is the link mass, l is the link length, and K is the stiffness of the joint.

To estimate the states of this system a three-layer neural network is applied which has 5 neurons in the input layer, 5 neurons in the hidden layer, and 4 neurons in the output layer. u, $\hat{q}_1$, $\dot{\hat{q}}_1$, $\hat{q}_2$ and $\dot{\hat{q}}_2$ are considered as input of the neural network. The input layer neurons have tangent hyperbolic transfer functions and the output neurons use linear activation functions. The initial weights of the network are selected as small random numbers.

Fig. 2.3 depicts the state estimation obtained by our proposed neural network. Fig. 2.3–a illustrate the responses of q_2 and $\hat{q}_2$ and Fig. 2.3–b shows those of q_1 and $\hat{q}_1$. Fig. 2.3–c and 2.3–d depict the responses of the motor and link velocities, respectively. It is clear that the states of the observer follow the states of the actual system.

2.6.1.1 Off-line Training:

In the next step, to evaluate the performance of the neural network when it is working as an off-line training scheme the training is stopped and the weights obtained from the last simulation results are considered as initial weights. Therefore, the network is used in recall mode. Fig. 2.4 illustrates the estimated stats when the neural network is trained off-line, i.e., the weights are not updated. These results demonstrate that the neural network has learned the system dynamics very accurately and can estimate the stats after training, properly.

2.6.1.2 Generalization

To investigate the generalization issue, another simulation is performed on the system. Toward this end, the training trajectory are defined as $0.1sint + 0.2sin2t + 0.05sin4t$. The results are given in Fig. 2.5–a to 2.5–e. Fig. 2.5–a depicts the state response of q_2 and $\hat{q}_2$ for the closed-loop system and Fig. 2.5–b illustrates those for q_1 and $\hat{q}_1$. Figs. 2.5–c to 2.5–e depict the state responses of the system in a short interval after the learning.

Later, we stop the training and applied a totally different trajectory namely $0.075sin3t$ to the manipulator. The simulation results obtained in this case are shown in Figs. 2.6–a to 2.6–d. Fig. 2.6–a depicts the state response of q_2 and $\hat{q}_2$ for the closed-loop system and Fig. 2.6–b shows those for q_1 and $\hat{q}_1$. Figs. 2.6–c and 2.6–d illustrate the responses for the joint and link velocities and their estimations. As can be observed, the proposed neural observer exhibits the desired generalization property.

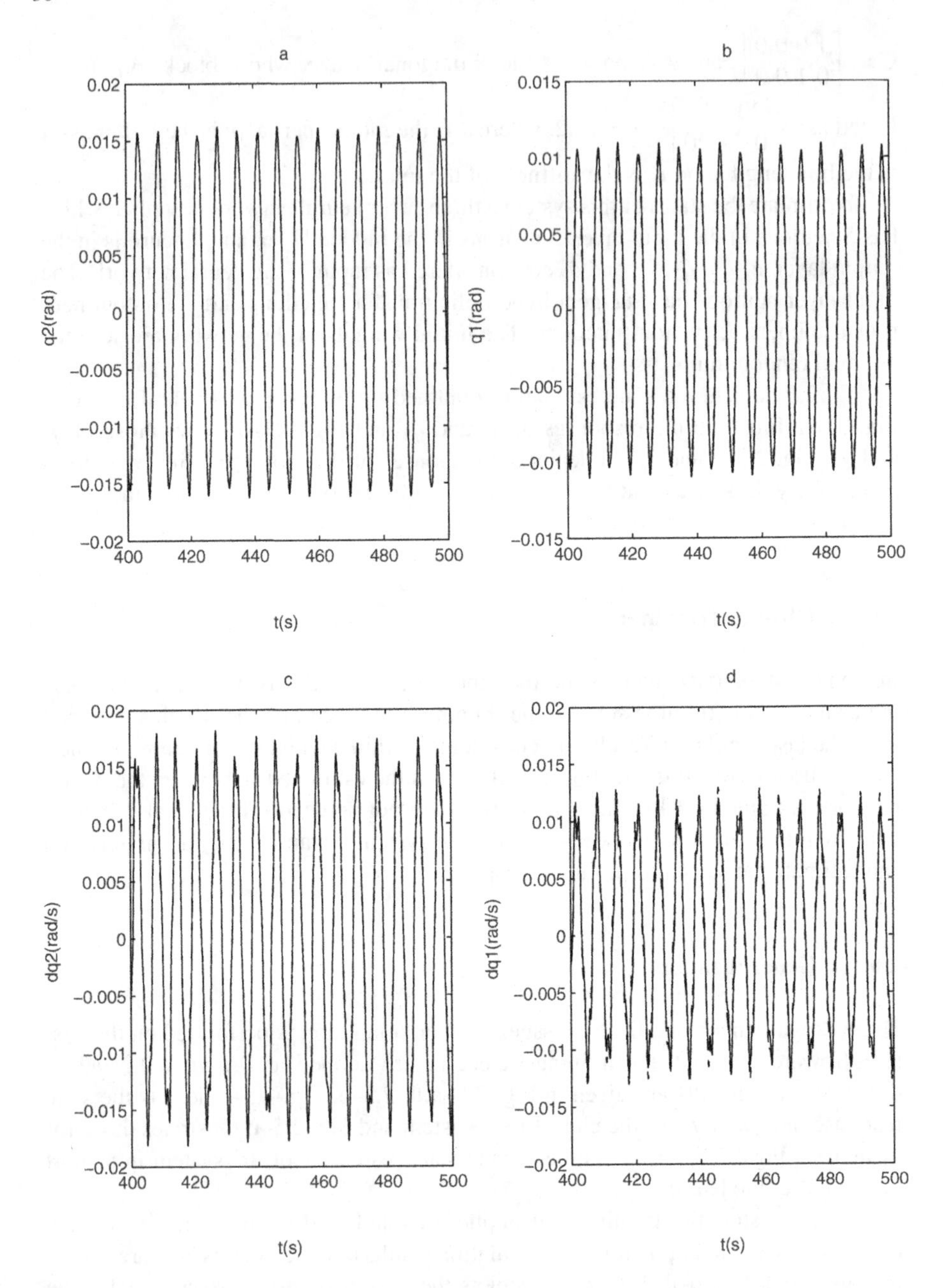

Fig. 2.3 The state responses of the single-link flexible-joint manipulator to $\sin(t)$ reference trajectory for NLPNN: (a) motor position, (b) link position, (c) motor velocity, (d) link velocity. The solid lines correspond to the actual states and the dashed lines correspond to the states of the observer.

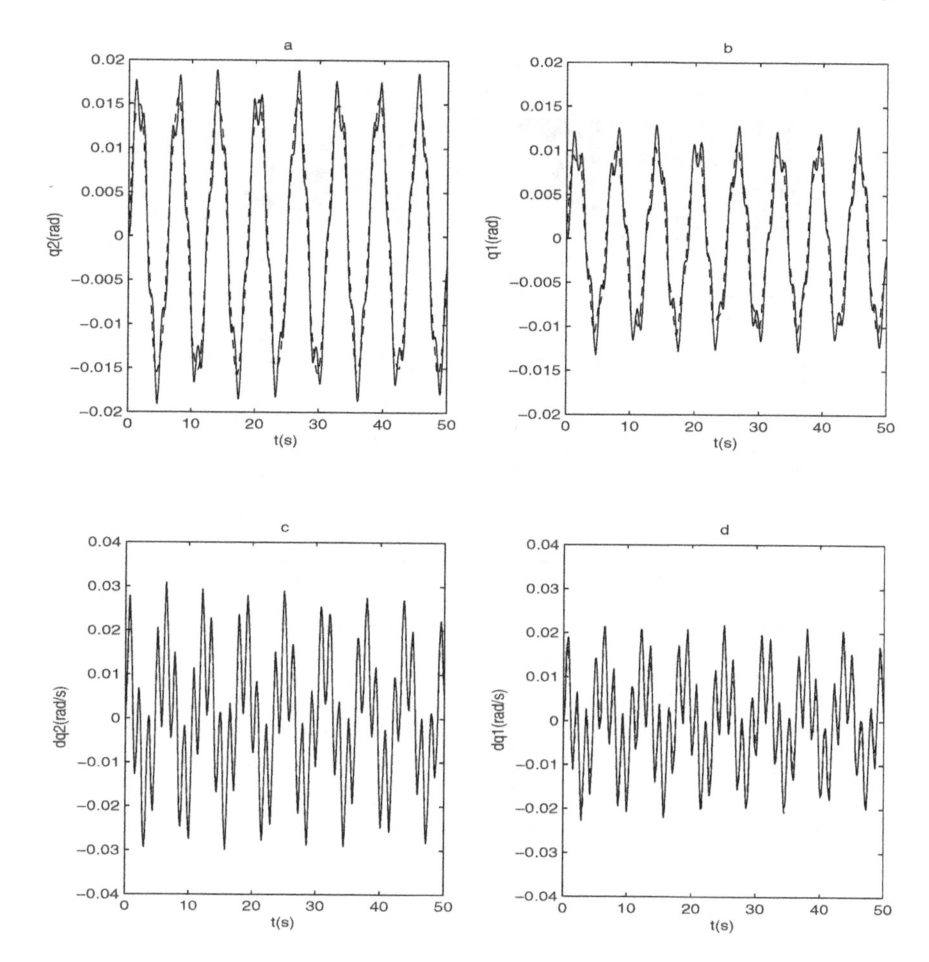

Fig. 2.4 The state responses of the single-link flexible-joint manipulator to $\sin(t)$ reference trajectory for NLPNN after the learning stops: (a) motor position, (b) link position, (c) motor velocity, (d) link velocity. The solid lines correspond to the actual states and the dashed lines correspond to the states of the observer.

2.6.2 A Two-link Flexible-Joint Manipulator

Simulation results for a two–link planar manipulator are presented in this section. The dynamics of a two-link manipulator are far more complicated than those of a single-link manipulator. The manipulator consists of two flexible-joints with state vector $x = [x_1\ x_2]$ where $x_i = [q_{1i}\ \dot{q}_{1i}\ q_{2i}\ \dot{q}_{2i}]$ is the state vector of ith link for $i = 1,2$ and the following numerical data

$$J = diag\{1.16, 1.16\},\ m = diag\{1, 1\},\ l_1 = l_2 = 1m,$$
$$K = diag\{100, 100\},\ \eta_1 = \eta_2 = 100, \text{ and } \rho_1 = \rho_2 = 1.5,$$

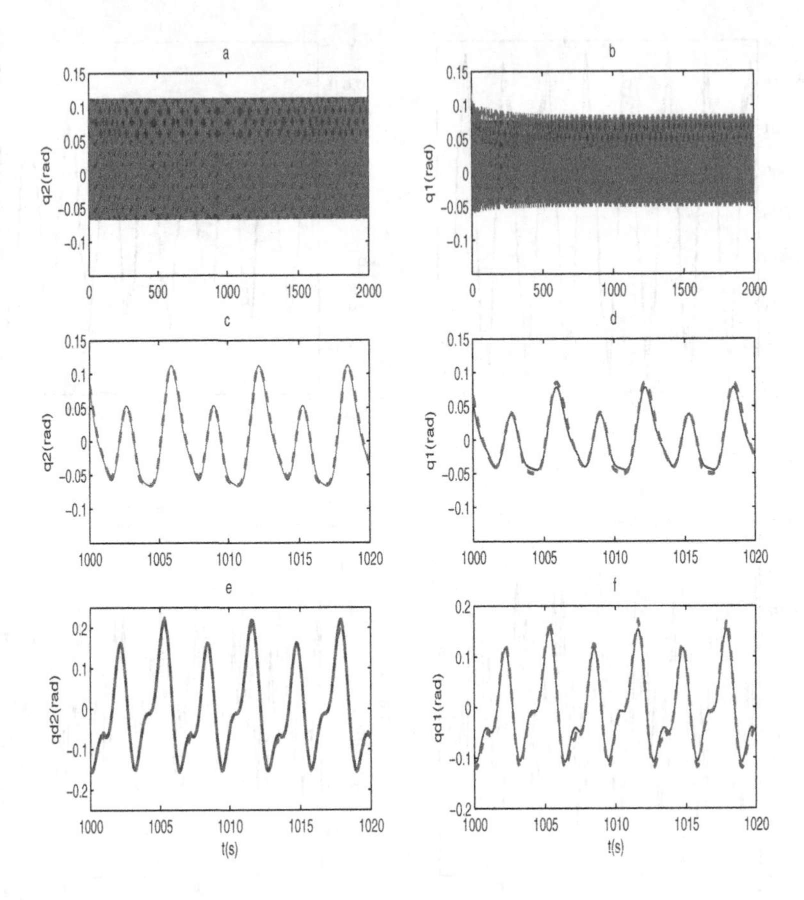

Fig. 2.5 The state responses of the single-link flexible-joint manipulator to $0.1sint + 0.2sin2t + 0.05sin4t$ reference trajectory for NLPNN: (a) motor position during the learning, (b) link position during the learning, (c) motor position at the end of learning phase, (d) link position at the end of learning phase, (e) motor velocity, (f) link velocity. The solid lines correspond to the actual states and the dashed lines correspond to the states of the observer.

and A is an 8×8 block diagonal matrix whose blocks A_{ii} are selected as $\begin{bmatrix} -20 & 1 \\ 0 & -20 \end{bmatrix}$, $i = 1 \ldots 4$. The neural network has three layers including 10 neurons in input layer, 10 neurons with tangent hyperbolic activation functions in hidden layer, and the 8 neurons with linear transfer functions in output layer. The input of the network is $\hat{\hat{x}}$. Fig. 2.7 depicts the result of the state estimation. As can be observed, despite the increased complexity in the manipulator model, the neural network has learned the manipulator dynamics, and all the states of the neural network track the corresponding states of the system. In the last step, for the sake of comparison a simulation

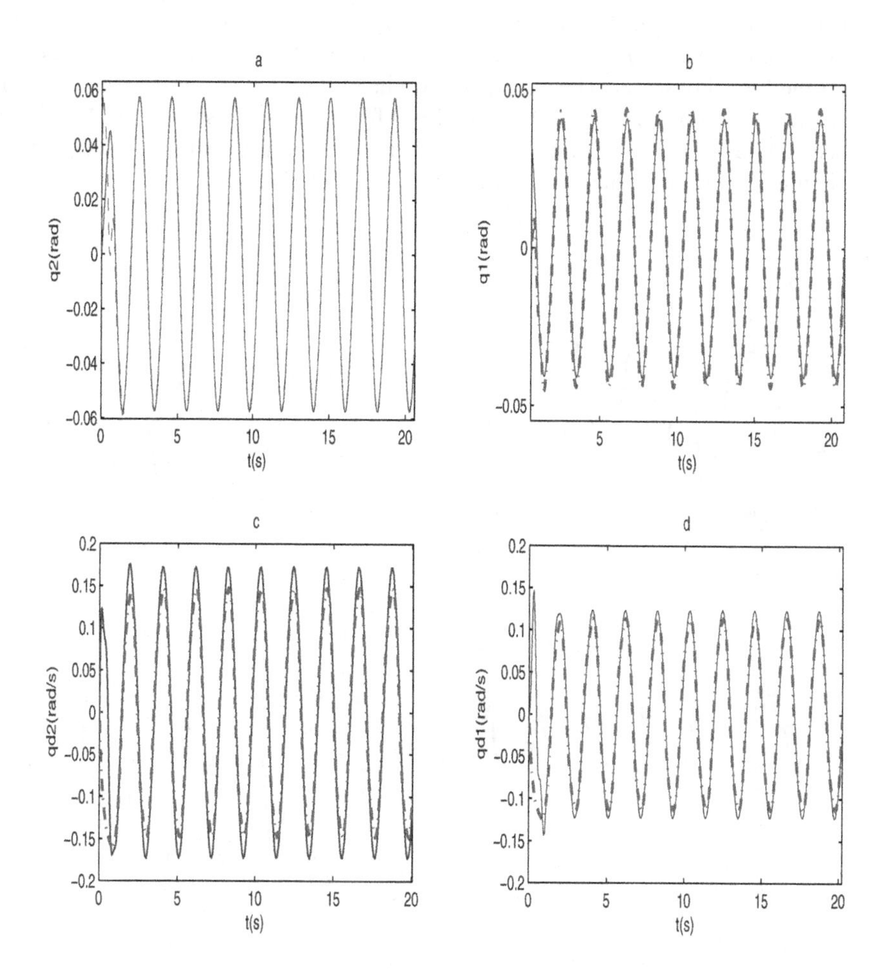

Fig. 2.6 The state responses of the single-link flexible-joint manipulator to $0.075 sin3t$ reference trajectory for NLPNN during the recall (testing) phase: (a) motor position, (b) link position, (c) motor velocity, (d) link velocity. The solid lines correspond to the actual states and the dashed lines correspond to the states of the observer.

is performed using LPNN and NLPNN in the same situation. Fig. 2.8–a illustrates the state responses, q_2 and $\hat{q}_2$ and Fig. 2.8–c shows the state responses, q_1 and $\hat{q}_1$ for the LPNN observer. The simulation results of the NLPNN observer in the same situation are shown in Figs. 2.8–b and 2.8–d for q_2 and q_1, respectively. As it is expected, the figures confirm that the obtained results by using the NLPNN observer are much superior to those obtained using the LPNN observer.

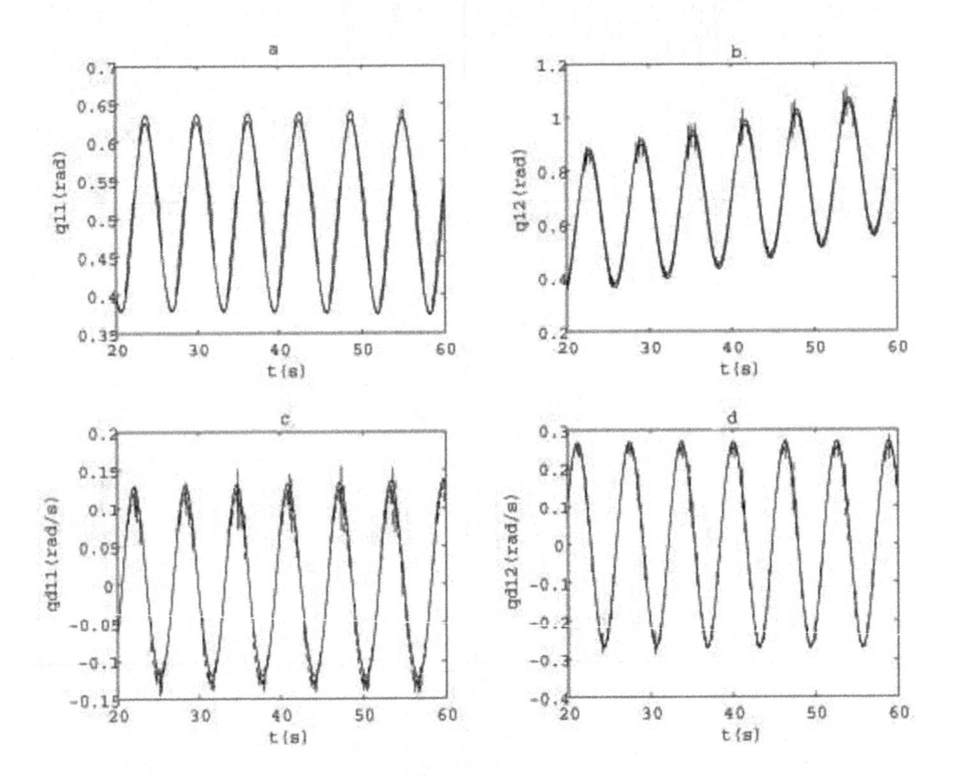

Fig. 2.7 The state responses of the two-link flexible-joint manipulator (after the learning period) to $sin(t)$ reference trajectory for NLPNN: (a) The position of the first link, (b) the position of the second link, (c) the velocity of the first link, (d) the velocity of the second link. The solid lines correspond to the actual states and the dashed lines correspond to the states of the observer.

2.7 Conclusions

Two recurrent neuro-adaptive observers for a general model of MIMO nonlinear systems have been introduced in this chapter. The structure of the proposed stable observers are considered to be either *linear* or *nonlinear* in parameters. The neural network weights have been updated based on the combination of a modified back-

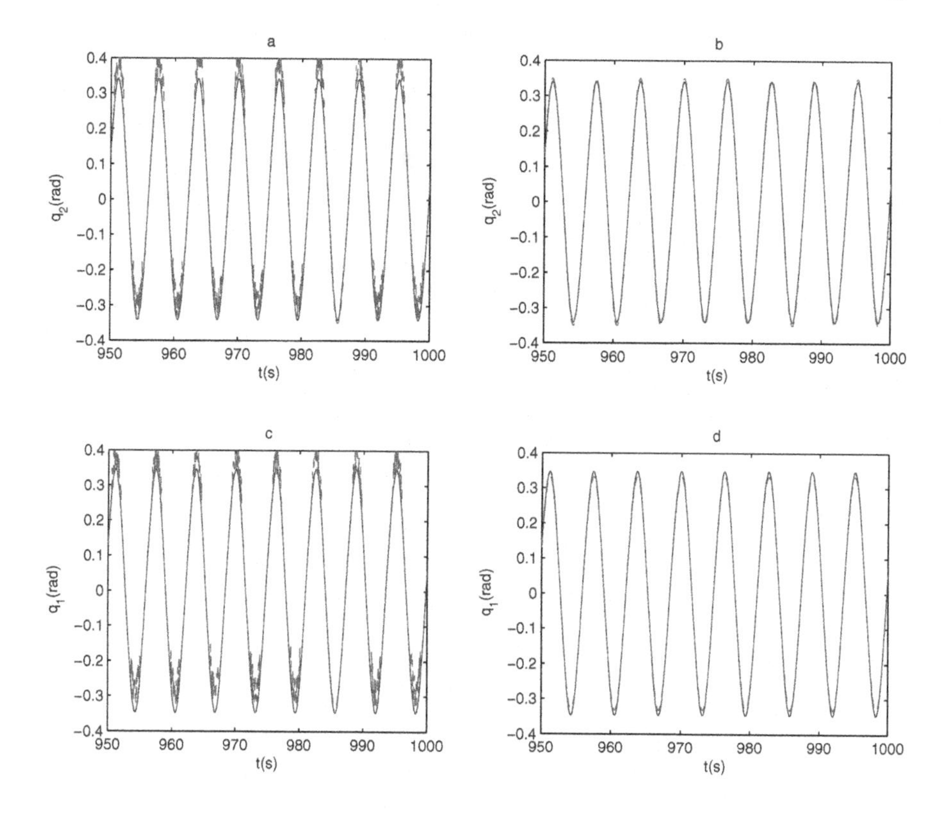

Fig. 2.8 The state responses of the flexible-joint manipulator (after the learning period) to $\sin(t)$ reference trajectory : (a) motor position for LPNN, (b) motor position for NLPNN, (c) link position for LPNN, (d) link position for NLPNN. The solid lines correspond to the actual states and the dashed lines correspond to the states of the observer.

propagation algorithm and an e-modification that guarantees the boundedness of the state estimation error. The stability of the overall system was shown by Lyapunov's direct method. It is worth noting that no SPR assumption or any other constraints that restrict the applicability of the approach was imposed on the system. The proposed observer can be applied both as an online and an off-line estimator. Simulation results performed on a flexible-joint manipulator confirm the reliable performance of the proposed observer.

Chapter 3
Neural Network-Based System Identification Schemes

3.1 Introduction

System identification is an important issue in determining a dynamical model for an unknown plant as well as in monitoring and control of system states. This field of study has been the focus of attention of researchers for several years. Online system identification to date is mostly based on recursive methods such as least squares [73].

Neural Networks have been widely employed for system identification since they can learn complex mappings from a set of examples. The mapping property, the adaptive nature and the ability of neural networks to deal with uncertainties make them viable choices for identification of nonlinear systems. There are several criteria that can be used to categorize neural network-based identifiers for nonlinear systems. For instance, one can categorize them via their input-state-output representations. There are then two basic categories: state-space representations and input-output representations. If the states of the system are available, the state-space representation allows the user to employ static neural networks. Using an input-output representation, however, needs some type of dynamic elements. Tap delay line and recurrent neural networks are commonly used to model dynamical systems.

In [74]-[77], several approaches were developed for identification and control of nonlinear systems based on stability theory. However, either some *a priori* knowledge about the nonlinear system is assumed or their adaptation laws are too complicated therefore in practice they are only applicable to certain classes of nonlinear systems. Such assumptions are often not valid for flexible-link manipulators such as Space Station Remote Manipulator System (SSRMS). In [78], a recurrent fuzzy neural network was presented for identification and control of dynamical systems. Mahdavi *et. al.* [79] experimentally verified an output feedback neural controller for DC-DC converter. However, no mathematical proof of stability was provided. On the other hand, backpropagation is widely used in classification and identification and it has been shown to give promising results, e.g. [79, 56, 55]. However, the main

H.A. Talebi et al., *Neural Network-Based State Estimation of Nonlinear Systems*, Lecture Notes in Control and Information Sciences 395,
DOI 10.1007/978-1-4419-1438-5_3, © Springer Science+Business Media, LLC 2010

drawback of the mentioned work on backpropagation is the lack of a mathematical proof of stability.

In this chapter, the neural network designed for observer in the previous chapter is modified for identification of general MIMO nonlinear systems. Unlike many other methods, the proposed approach does not assume knowledge of the nonlinearities of the system nor that the nonlinear system is linear in its parameters. Both *parallel* and *series-parallel* models are considered. As a case study, identification of the dynamics of flexible-link manipulators are considered to demonstrate the excellent performance of the proposed schemes.

The reminder of this chapter is organized as follows. In Section 3.2, two neural network identification schemes for parallel model are proposed. At first, a linearly parameterized neural network (LPNN) is introduced in Section 3.2.1 and also a mathematical proof of stability is given. The results of Section 3.2.1 are extended to a more general model of MIMO nonlinear systems in Section 3.2.2. In this section, a *nonlinear-in-parameters* neural network (NLPNN) is introduced that takes advantage of the full capability of universal approximation theory. The proposed identifier then modified for series-parallel identifier model in Section 3.3. Section 3.4 presents a case study on the: identification of the dynamics of a flexible-link manipulator. Section 3.5 gives some simulation results. The proposed neuro-identification scheme has been implemented on an experimental set-up consisting of a three-link macro-micro manipulator. Section 3.6 describes the test-bed and demonstrates the experimental results. Finally, Section 3.7 gives some conclusions.

3.2 A Parallel Identification Scheme

In this section, the LPNN and NLPNN introduced in the previous chapter are employed to identify the dynamics of an unknown nonlinear system given by

$$\dot{x} = f(x,u), \tag{3.1}$$

where $u \in R^m$ is the input vector and $x \in R^n$ is the state vector of the system and $f(.)$ is an unknown nonlinear function. Similar to the previous chapter, it should be assumed that the open loop system (3.1) is stable.

By following along the similar steps given in the previous chapter, i.e., Ax is added to and subtracted from (3.1), where A is an arbitrary Hurwitz matrix, we can get

$$\dot{x} = Ax + g(x,u), \tag{3.2}$$

where $g(x,u) = f(x,u) - Ax$. Based on (3.2), a recurrent network model can be constructed by parameterizing the mapping g by feedforward (static) neural network architectures, denoted by N. Therefore, the following model is considered for identification purposes.

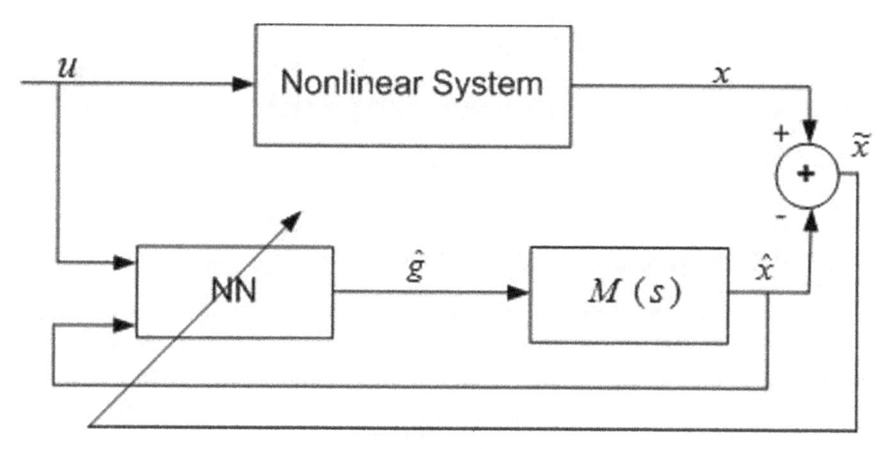

Fig. 3.1 The structure of neural network identifier (parallel model).

$$\dot{\hat{x}} = A\hat{x} + \hat{g}(\hat{x}, u), \tag{3.3}$$

where $\hat{x}$ denotes the estimated states. To identify the nonlinear dynamics, in this section the so-called parallel identifier is considered which means that only the estimation error of the states are available to identifier, i.e., the states of the estimated model are fed to the input of the neural network. The structure of the parallel identifier is shown in Fig. 3.1. Corresponding to the Hurwitz matrix A, $M(s) := (sI - A)^{-1}$ is also shown which is an $n \times n$ matrix whose elements are stable transfer functions.

In the light of Theorem 1.1, a multilayer NN can be designed to estimate the nonlinear function $g(x,u)$ by the following model

$$g(x,u) = W\sigma(V\bar{x}) + \varepsilon(x), \tag{3.4}$$

where W and V are the ideal but unknown weight matrices and $\bar{x} = [x \;\; u]^T$, $\varepsilon(x) \leq \varepsilon_N$ is the neural network's bounded approximation error, and $\sigma(.)$ is the transfer function of the hidden neurons considered as tangent hyperbolic function represented in (2.4).

As far as the structure of the neural network is concerned, by comparing the neuro-observer schematic in Fig. 2.1 with the parallel identifier scheme in Fig. 3.1, it can be seen that even though both receive the same inputs, the neuro-identifier is trained based on the stat estimation error. Whereas, in neuro-observer since all states are not available output error is used to train the network.

In the following two subsections a stable LPNN as well as a NLPNN neuro-identifier are introduced.

3.2.1 LPNN Parallel Identifier

To design an LPNN identifier, let us substitute $V = I$ in (3.4). Hence, the function g can be expressed as

$$g(x,u) = W\sigma(\bar{x}) + \varepsilon(x). \tag{3.5}$$

The function g can be approximated by an LPNN as

$$\hat{g}(\hat{x},u) = \hat{W}\sigma(\hat{\bar{x}}). \tag{3.6}$$

Therefore, the proposed identifier is given by

$$\dot{\hat{x}}(t) = A\hat{x} + \hat{W}\sigma(\hat{\bar{x}}). \tag{3.7}$$

Defining the identification error as $\tilde{x} = x - \hat{x}$ and by using (3.2), (3.5) and (3.7), the error dynamics can be expressed as

$$\dot{\tilde{x}}(t) = A\tilde{x} + \tilde{W}\sigma(\hat{\bar{x}}) + w(t) + \varepsilon(x), \tag{3.8}$$

where $\tilde{W} = W - \hat{W}$, $w(t) = W[\sigma(\bar{x}) - \sigma(\hat{\bar{x}})]$ is a bounded disturbance term i.e., $\|w(t)\| \leq \bar{w}$ for some positive constant $\bar{w}$, due to the tangent hyperbolic function.

In the following theorem, the learning rule introduced in Theorem 2.1 for updating the weights is modified for parallel identifier and the stability of the overall system is shown by Lyapunov's direct method.

Theorem 3.1. *Consider the plant model (3.1) and the identifier model (3.7). Given Assumption 2.2, if the weights of the LPNN are updated according to*

$$\dot{\hat{W}} = -\eta(\frac{\partial J}{\partial \hat{W}}) - \rho\|\tilde{x}\|\hat{W}, \tag{3.9}$$

where $\eta > 0$ is the learning rate, $J = \frac{1}{2}(\tilde{x}^T\tilde{x})$ is the objective function and ρ is a small positive number, then $\tilde{x}$ and $\tilde{W} \in L_\infty$. In other words, the identification and weight errors are ultimately bounded.

Proof: First, note that $\frac{\partial J}{\partial \hat{W}}$ can be computed according to

$$\frac{\partial J}{\partial \hat{W}} = \frac{\partial J}{\partial \tilde{x}}\frac{\partial \tilde{x}}{\partial \hat{x}}\frac{\partial \hat{x}}{\partial \hat{W}} = -\tilde{x}^T\frac{\partial \hat{x}}{\partial \hat{W}}. \tag{3.10}$$

Now, by using the static approximation of $\frac{\partial \hat{x}}{\partial \hat{W}}$ as $-A^{-1}\frac{\partial \hat{g}}{\partial \hat{W}}$ and by using (3.6) and (3.8), the learning rule (3.9) in terms of $\tilde{W}$ may be written as

$$\dot{\tilde{W}} = \eta(\tilde{x}^T A^{-1})^T(\sigma(\hat{\bar{x}}))^T + \rho\|\tilde{x}\|\hat{W}. \tag{3.11}$$

Consider the Lyapunov function candidate

$$L = \frac{1}{2}\tilde{x}^T P\tilde{x} + \frac{1}{2}tr(\tilde{W}^T\rho^{-1}\tilde{W}), \tag{3.12}$$

where $P = P^T > 0$. Then, the time derivative of (3.12) is given by

$$\dot{L} = \frac{1}{2}\dot{\tilde{x}}^T P\tilde{x} + \frac{1}{2}\tilde{x}^T P\dot{\tilde{x}} + tr(\tilde{W}^T\rho^{-1}\dot{\tilde{W}}). \tag{3.13}$$

Now, by using the Lyapunov equation $A^T P + PA = -Q$ for a Hurwitz matrix A and a symmetric positive definite matrix Q, and substituting (3.8) and (3.11) into (3.13), one gets

$$\begin{aligned}\dot{L} =& -\frac{1}{2}\tilde{x}^T Q\tilde{x} + \tilde{x}^T P(\tilde{W}\sigma(\hat{\tilde{x}}) + w + \varepsilon)\\ &+ tr(-\tilde{W}^T l\tilde{x}\sigma^T + \tilde{W}^T\|\tilde{x}\|(W - \tilde{W})),\end{aligned} \tag{3.14}$$

where $l = \eta\rho^{-1}A^{-T}$. Moreover, we have

$$tr(\tilde{W}^T(W - \tilde{W}) \leq W_M\|\tilde{W}\| - \|\tilde{W}\|^2 \tag{3.15}$$

$$tr(\tilde{W}^T l\tilde{x}\sigma^T) \leq \sigma_m\|\tilde{W}^T\|\ \|l\|\ \|\tilde{x}\|, \tag{3.16}$$

where $\|W\|| \leq W_M$ and $\|\sigma(\hat{\tilde{x}})\| \leq \sigma_m$. Now by using (3.15) and (3.16), (3.14) can be written as

$$\begin{aligned}\dot{L} \leq& -\frac{1}{2}\lambda_{min}(Q)\|\tilde{x}\|^2 + \|\tilde{x}\|\|P\|(\|\tilde{W}\|\sigma_m + \bar{w} + \varepsilon_N)\\ &+ \sigma_m\|\tilde{W}\|\|l\|\|\tilde{x}\| + (W_M\|\tilde{W}\| - \|\tilde{W}\|^2)\|\tilde{x}\|.\end{aligned} \tag{3.17}$$

Furthermore, by completing the squares in (3.17) the following inequality guarantees negative semidefiniteness of $\dot{L}$ and therefore, ultimate boundedness of $\tilde{x}$.

$$\|\tilde{x}\| \geq \frac{(2\|P\|(\bar{w} + \varepsilon_N) + (\sigma_m\|P\| + W_M + \sigma_m\|l\|)^2/2)}{\lambda_{min}(Q)} = b. \tag{3.18}$$

Following along similar lines given in proof of Theorem 2.1, it can be assured that by using the given updating law, (3.9) the estimation error of weight matrix W, i.e., $\|\tilde{W}\|$ is bounded as well. This completes the proof of the theorem. □

By applying NLPNN structure in the next section, a parallel identifier for a larger class of nonlinear systems is presented.

3.2.2 *NLPNN Parallel Identifier*

Consider the nonlinear system (3.1) and identifier model (3.3) and suppose that the assumption concerning the stability of the open-loop system (3.1) holds as well. In this case, the function g can be approximated as

$$\hat{g}(\hat{x}, u) = \hat{W}\sigma(\hat{V}\hat{\bar{x}}). \tag{3.19}$$

The proposed identifier can now be given by

$$\dot{\hat{x}}(t) = A\hat{x} + \hat{W}\sigma(\hat{V}\hat{\bar{x}}). \tag{3.20}$$

Defining the identification error as $\tilde{x} = x - \hat{x}$, and using (3.2), (3.19) and (3.20), the error dynamics can be stated as

$$\dot{\tilde{x}}(t) = A\tilde{x} + \tilde{W}\sigma(\hat{V}\hat{\bar{x}}) + w(t), \tag{3.21}$$

where $\tilde{W} = W - \hat{W}$, $w(t) = W[\sigma(V\bar{x}) - \sigma(\hat{V}\hat{\bar{x}})] + \varepsilon(x)$ is a bounded disturbance term, i.e., $\|w(t)\| \leq \bar{w}$ for some positive constant $\bar{w}$, due to the tangent hyperbolic function.

It can be seen that $\hat{V}$ appears in (3.21) only as an argument of a tangent hyperbolic function which is known to be bounded. Hence, the first step in decomposing the system (which greatly simplifies the stability analysis as will be shown later) can be initiated by introducing a bounded variable S_1 defined below

$$S_1 = \sigma(\hat{V}\hat{\bar{x}}). \tag{3.22}$$

Equation (3.21) can then be written as

$$\dot{\tilde{x}}(t) = A_c\tilde{x} + \tilde{W}S_1 + w(t). \tag{3.23}$$

In Theorem 3.1, we modify the BP algorithm for a *nonlinear-in-parameters* neural network (NLPNN) and then present a new approach for stability analysis of the proposed scheme. The proposed stability approach is based on decomposing the neural network into two subsystems. The first subsystem consists of the estimation and output-layer weight error ($\tilde{x}$ and $\tilde{W}$) and the second subsystem consists of the hidden layer weight error ($\tilde{V}$). The key to this decomposition is that the hidden layer weights appear in the first subsystem, only as an argument of a tangent hyperbolic function S_1 and its derivative S_2 which are both known to be bounded. This allows us to regard the first subsystem as a linear-in-parameter neural network (LPNN) whose stability was shown in the previous section. Having shown the stability of the first subsystem, the stability of the second subsystem is shown subsequently without the requirement of having the limiting assumptions mentioned in [80].

Theorem 3.2. *Consider the plant model (3.1) and the identifier model (3.20). Given Assumption 2.2, if the weights of NLPNN are updated according to:*

$$\dot{\hat{W}} = -\eta_1\left(\frac{\partial J}{\partial \hat{W}}\right) - \rho_1\|\tilde{x}\|\hat{W} \tag{3.24}$$

$$\dot{\hat{V}} = -\eta_2\left(\frac{\partial J}{\partial \hat{V}}\right) - \rho_2\|\tilde{x}\|\hat{V}, \tag{3.25}$$

then $\tilde{x}$, $\tilde{W}$, and $\tilde{V} \in L_\infty$, i.e., the estimation error and the weights error are all ultimately bounded. In these equations, η_1, $\eta_2 > 0$ are the learning rates, ρ_1, ρ_2 are some small positive numbers, and $J = \frac{1}{2}(\tilde{x}^T\tilde{x})$ is the neural network objective function.

The first terms in (3.24) and (3.25) are the popular backpropagation algorithm and the last terms are the e-modification terms which add extra damping for robustness.

Proof: Let us define:

$$net_{\hat{v}} = \hat{V}\hat{\bar{x}} \tag{3.26}$$
$$net_{\hat{w}} = \hat{W}\sigma(\hat{V}\hat{\bar{x}}). \tag{3.27}$$

Therefore, by using the chain rule $\frac{\partial J}{\partial \hat{W}}$ and $\frac{\partial J}{\partial \hat{V}}$ can be computed according to

$$\frac{\partial J}{\partial \hat{W}} = \frac{\partial J}{\partial net_{\hat{w}}} \cdot \frac{\partial net_{\hat{w}}}{\partial \hat{W}}$$
$$\frac{\partial J}{\partial \hat{V}} = \frac{\partial J}{\partial net_{\hat{v}}} \cdot \frac{\partial net_{\hat{v}}}{\partial \hat{V}},$$

where

$$\frac{\partial J}{\partial net_{\hat{w}}} = \frac{\partial J}{\partial \tilde{x}} \cdot \frac{\partial \tilde{x}}{\partial \hat{x}} \cdot \frac{\partial \hat{x}}{\partial net_{\hat{w}}} - \tilde{x}^T \cdot \frac{\partial \hat{x}}{\partial net_{\hat{w}}}$$
$$\frac{\partial J}{\partial net_{\hat{v}}} = \frac{\partial J}{\partial \tilde{x}} \cdot \frac{\partial \tilde{x}}{\partial \hat{x}} \cdot \frac{\partial \hat{x}}{\partial net_{\hat{v}}} - \tilde{x}^T \cdot \frac{\partial \hat{x}}{\partial net_{\hat{v}}}, \tag{3.28}$$

and

$$\frac{\partial net_{\hat{w}}}{\partial \hat{W}} = \sigma(\hat{V}\hat{\bar{x}})$$
$$\frac{\partial net_{\hat{v}}}{\partial \hat{V}} = \hat{\bar{x}}. \tag{3.29}$$

We modify the original BP algorithm such that the static approximations of $\frac{\partial \hat{x}}{\partial net_{\hat{w}}}$ and $\frac{\partial \hat{x}}{\partial net_{\hat{v}}}$ ($\dot{\hat{x}} = 0$) can be used. Thus, using (3.20), (3.26), and (3.27), we can write

$$\frac{\partial \hat{x}}{\partial net_{\hat{w}}} \approx -A^{-1}$$
$$\frac{\partial \hat{x}}{\partial net_{\hat{v}}} \approx -A^{-1}\hat{W}(I - \Lambda(\hat{V}\hat{\bar{x}})), \tag{3.30}$$

where

$$\Lambda(\hat{V}\hat{\bar{x}}) = diag\{\sigma_i^2(\hat{V}_i\hat{\bar{x}})\}, i = 1, 2, ..., m. \tag{3.31}$$

Now, by substituting (3.28), (3.29), and (3.30) in (3.24) and (3.25), we have

$$\dot{\hat{W}} = -\eta_1(\tilde{x}^T A^{-1})^T(\sigma(\hat{V}\hat{\bar{x}}))^T - \rho_1\|\tilde{x}\|\hat{W} \tag{3.32}$$
$$\dot{\hat{V}} = -\eta_2(\tilde{x}^T A^{-1}\hat{W}(I - \Lambda(\hat{V}\hat{\bar{x}})))^T\hat{\bar{x}}^T - \rho_2\|\tilde{x}\|\hat{V}. \tag{3.33}$$

Given $\tilde{W} = W - \hat{W}$ and $\tilde{V} = V - \hat{V}$, where W and V are the fixed ideal weights, the weight error dynamics can be described by:

$$\dot{\tilde{W}} = \eta_1(\tilde{x}^T A^{-1})^T(\sigma(\hat{V}\hat{\bar{x}}))^T + \rho_1\|\tilde{x}\|\hat{W} \tag{3.34}$$
$$\dot{\tilde{V}} = \eta_2(\tilde{x}^T A^{-1}\hat{W}(I - \Lambda(\hat{V}\hat{\bar{x}})))^T\hat{\bar{x}}^T + \rho_2\|\tilde{x}\|\hat{V}. \tag{3.35}$$

Now, we introduce another variable S_2 as

$$S_2 = I - \Lambda(\hat{V}\hat{\bar{x}}). \tag{3.36}$$

Note that S_2 is in fact the derivative of the tangent hyperbolic function which is also known to be bounded. Next, we define the learning rules (3.34) and (3.35) in terms of the new variables S_1 and S_2 as

$$\dot{\tilde{W}} = l_0\tilde{x}S_1^T + \rho_1\|\tilde{x}\|\hat{W} \tag{3.37}$$
$$\dot{\tilde{V}} = S_2^T\hat{W}^T l_2\tilde{x}\hat{\bar{x}}^T + \rho_2\|\tilde{x}\|\hat{V}, \tag{3.38}$$

where S_1 and S_2 are given by (3.22) and (3.36), respectively; $l_0 = \eta_1 A^{-T}$; and $l_2 = \eta_2 A^{-T}$. Now, the following observations can be made:

Fact 3.2.1 *The boundedness of $\tilde{V}$ has no effect on the stability of $\tilde{x}$ (3.23) and $\tilde{W}$ (3.37) since $S_1 \in L_\infty$ regardless of the value of $\tilde{V}$.*

Fact 3.2.2 *The function $\Lambda(.)$ defined in (3.31) is bounded due to the boundedness of tangent hyperbolic function, hence $S_2 \in L_\infty$.*

Given the Fact 3.2.1, we can conclude that the system can be decomposed into two subsystems for the purpose of stability analysis. One subsystem (Subsystem 1) consists of the estimation error dynamics (3.23) and output-layer weight error (3.37) and the other subsystem (Subsystem 2) consists of the hidden layer weight error (3.38). Note that, the two subsystems are not completely decoupled due to the presence of $\tilde{x}$ and $\hat{W}$ in (3.38). Consequently, the stability of Subsystem 1 can be shown independently of Subsystem 2, but not vice versa. Now, consider S_1 as the input to Subsystem 1. The boundedness of S_1 will allows us to treat this subsystem as a linear-in-parameter neural network. The stability of such systems was shown in the previous section. Hence, by adopting the same approach as in LPNN case, the ultimate boundedness of the estimation error $\tilde{x}$ and the output layer weight error $\tilde{W}$ is guaranteed.

Since the open-loop system is stable and the ideal weights are also constant, we can conclude that $\hat{W} \in L_\infty$ and $\hat{x} \in L_\infty$.

To show the boundedness of Subsystem 2 (hidden layer weight error $\tilde{V}$), consider (3.38) which can be rewritten as

$$\dot{\tilde{V}} = f_2(\tilde{x}, \hat{W}, \hat{V}) + \rho_2\|\tilde{x}\|\hat{V}$$
$$= f_2(\tilde{x}, \hat{W}, \hat{V}) + \alpha V - \alpha\tilde{V}, \tag{3.39}$$

where

$$f_2(\tilde{x}, \hat{W}, \hat{V}) = S_2^T \hat{W}^T l_2 \tilde{x}\hat{\bar{x}}^T$$
$$\alpha = \rho_2\|\tilde{x}\|.$$

It can be seen that $f_2(.)$ is bounded since $\tilde{x}$, $\hat{\bar{x}}$, $\hat{W}$ and S_2 are all bounded, and A is a Hurwitz matrix. Given the fact the the ideal weight V is fixed, (3.39) can be regarded as a linear system with bounded input $(f_2(\tilde{x}, \hat{W}, \hat{V}) + \alpha V)$. It is clear that this system is stable since α is positive and the system input remains bounded. Hence, the boundedness of $\tilde{V}$ is also ensured, i.e., $\tilde{V} \in L_\infty$. This completes the proof of Theorem 3.1. □

Remark 3.1 *The size of the estimation error bound b can be kept small by proper selection of the damping factors, the Hurwitz matrix A and the learning rates such that higher accuracy can be achieved. Similar to proposed observer in Chapter 2, for selecting a proper learning rate, there is a compromise between faster convergence and avoiding overshoot. Moreover, the Hurwitz matrix A has considerable effect on convergence as well as accuracy of the state estimation (see (3.34) and (3.35)). By selecting matrix A which its eigenvalues are farther to the left in the complex plane to provide a more stable matrix, due to applying A^{-1} in updating the weights it may decrease convergence rate of the weights to their ideal values. One suggestion can be providing better accuracy by selecting a more stable A and better convergence by using higher learning rates.*

Since in identification problems the states of the systems are mostly available, they can be applied for designing the neuor-identifier. In the following section, the introduced NLPNN identifier is modified for identification model in which the actual states rather than estimated states are fed to the neural network.

3.3 A Series-Parallel Identification Scheme

Taking advantage of the actual states instead of their estimations as the input to the neural networks provides another model of neuro-identifier namely, *series-parallel* model. It is well known that a *series-parallel* mode has better convergence properties than a *parallel* model [81]. Fig. 3.2 depicts a series-parallel neural network identifier. In this case, the function g can be approximated by NLPNN as

$$\hat{g}(x,u) = \hat{W}\sigma(\hat{V}\bar{x}). \tag{3.40}$$

In other words, as compared to (3.19), only $\hat{\bar{x}}$ is changed to $\bar{x}$. Therefore, the error dynamics can be expressed as

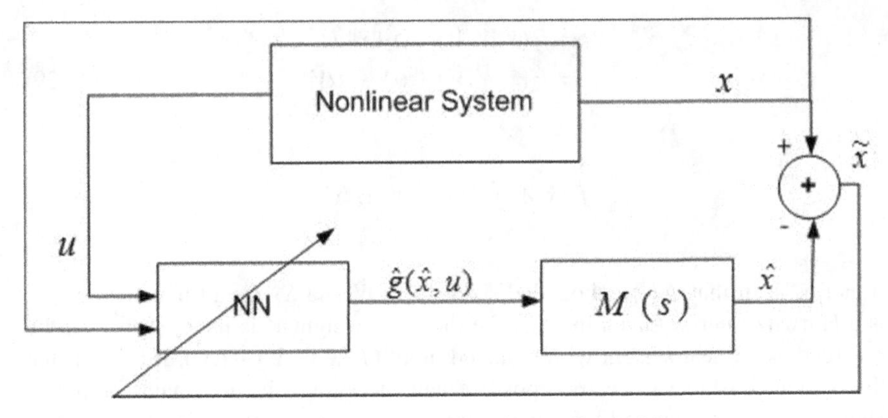

Fig. 3.2 The structure of neural network identifier (series-parallel model).

$$\dot{\tilde{x}}(t) = A\tilde{x} + \tilde{W}\sigma(\hat{V}\bar{x}) + w(t), \tag{3.41}$$

where $w(t) = W[\sigma(V\bar{x}) - \sigma(\hat{V}\bar{x})] + \varepsilon(x)$ and, (3.41) and (3.23) differ merely on the definition of $w(t)$. Applying this change to all other equations from (3.26) to (3.38), the rest of the stability proof remains unchanged.

In the next section, the flexible-link manipulator which is well-known as a highly nonlinear dynamics is considered as a case study to evaluate the performance of the proposed identifiers.

3.4 A Case study: Application to a Flexible-Link Manipulator

Structurally flexible robotic manipulators have been used in many applications. The basic motivation behind this field of research comes from several different sources. For instance, space applications require light-weight long-reach designs to accomplish tasks with low energy consumption. On the other hand, increased structural flexibility may be desirable in tasks such as cleaning delicate surfaces or avoiding damages to the manipulator system due to accidental collisions.

Structural flexibility, however, causes greater difficulty in modeling manipulator dynamics. This requires inclusion of deformation effects (due to the flexibility of the links) in the dynamic equations, which generally tends to complicate the analysis and design of the control laws. Flexible-link robot models belong to a class of distributed parameter systems described by partial differential equations (PDE's). The assumed mode and finite element methods are two common approaches currently used for approximating PDE's by a system of ordinary differential equations (ODE's). A relatively large number of flexible modes is required to accurately model the system behavior in this case.

Lagrangian and Newton-Euler formulations are two common approaches currently used for obtaining the flexible-link manipulator dynamics e.g. see [82, 83, 84]. However, unmodeled dynamics due to model truncation and presence of various friction and backlash terms always exist and contribute to the inaccuracy of the nominal model.

For the purpose of simulation, the dynamic model of a flexible-link manipulator is defined in the following form using the recursive Lagrangian approach:

$$M(q)\ddot{q}+C(q,\dot{q})\dot{q}+Kq+F\dot{q}=u, \tag{3.42}$$

where $u=[\tau^T\ 0_{1\times m}]^T$, $q=[\theta^T\ \delta^T]^T$, θ is the $n\times 1$ vector of joint variables, δ is the $m\times 1$ vector of deflection variables and $C=[C_1(q,\dot{q})\ C_2(q,\dot{q})]^T$ represents the terms due to gravity, Coriolis, and centrifugal forces; M is the mass matrix, $K=\begin{bmatrix}0_{n\times n} & 0_{n\times m}\\ 0_{m\times n} & K_{m\times m}\end{bmatrix}$ is the stiffness matrix, $F=diag\{F_1,F_2\}$ represents the viscous friction at the hub and in the structure, and τ is the input torque. It is common in identification schemes that the state $x(t)$ of the system is assumed to be bounded. In other words, the open-loop system is stable. This is due to the fact that we cannot expect the network to follow an unbounded signal. However, the flexible-link manipulator is an unstable or marginally stable open-loop system. In fact, the linearized system has one or two poles at the origin depending on the damping coefficients. It is proposed that a joint PD controller be added to the system. The joint PD control can stabilize the closed-loop system so that boundedness of the signal $x(t)$ can be assured. Consider the control law

$$u_c=-K_p\theta-K_v\dot{\theta}. \tag{3.43}$$

The following result is taken from [85, 71].

Theorem 3.3. *Assume that K_p and K_v are symmetric and positive-definite matrices. Then, the closed-loop system obtained by using (3.43) and (3.42) is asymptotically stable.*

Proof: The stability of the closed-loop system can be shown by selecting the following Lyapunov candidate function

$$V=\frac{1}{2}\dot{q}^T M(q)\dot{q}+\frac{1}{2}q^T(K+B^T K_p B)q, \tag{3.44}$$

where $B=[I_{n\times n}0_{n\times m}]$. Taking the time derivative of V yields in

$$\dot{V}=\dot{q}^T M\ddot{q}+\frac{1}{2}\dot{q}^T\dot{M}\dot{q}+\dot{q}^T\bar{K}q, \tag{3.45}$$

where $\bar{K}=K+B^T K_p B$. Now by substituting (3.42) in (3.45), we have,

$$\dot{V}=\dot{q}^T[B^T u-C\dot{q}-F\dot{q}-Kq]+\frac{1}{2}\dot{q}^T\dot{M}\dot{q}+\dot{q}^T\bar{K}q. \tag{3.46}$$

Now by using (3.43), (3.46) can be written as

$$\dot{V} = \dot{q}^T B^T(-K_p\theta - k_v\dot{\theta}) + \dot{q}^T(\frac{1}{2}\dot{M} - C)\dot{q} - \dot{q}^T F\dot{q} - \dot{q}^T Kq + \dot{q}^T \bar{K}q$$

$$\dot{V} = \dot{q}^T(\frac{1}{2}\dot{M} - C)\dot{q} - \dot{q}^T \bar{K}q + \dot{q}^T \bar{K}q - \dot{q}^T(F + B^T K_v B)\dot{q}. \quad (3.47)$$

It is well known that the matrix $(\dot{M} - 2C)$ is skew-symmetric [3] and hence $\dot{q}^T(\frac{1}{2}\dot{M} - C)\dot{q} = 0$, which results in

$$\dot{V} = -\dot{q}^T(F + B^T K_v B)\dot{q}. \quad (3.48)$$

Since $(F + B^T K_v B)$ is a positive definite symmetric matrix, it is clear that:

$$\dot{V} \leq 0, \quad (3.49)$$

i.e., V is negative semidefinite. Hence, V is decreasing as long as $\dot{q}$ is not zero. Now, Lassalle's theorem [32, 86] can be used to show asymptotic stability of the closed-loop system. Towards this end, suppose $\dot{V} \equiv 0$. Then, (3.48) implies that $\dot{q} \equiv 0$ and hence $\ddot{q} \equiv 0$. Substituting this result in closed-loop system equations, we get:

$$(F + B^T K_v B)q = 0 \implies q = 0.$$

Thus by using Lassalle's theorem, we can conclude that the system is globally asymptotically stable. For more details, refer to [85, 71].

Since the result of Theorem 3.3 is valid for any positive definite K_p and K_v, without using any *a priori* knowledge about the system dynamics, the control signal u_c defined by (3.43) can be added to the system to guarantee a stable closed-loop system. The next section provides simulations results for identifying dynamics of a two-link flexible manipulator as well as a Space Station Remote Manipulator System (SSRMS).

3.5 Simulation Results

In this section, the performance of the introduced identifier is evaluated by simulations results in three steps. In the first step, the nonlinear dynamics of a two-link flexible manipulator is identified by the proposed NLPNN parallel as well as series-parallel identifier. Then the introduced identifier is applied for identifying dynamics of SSRMS whose dynamics is highly nonlinear. In the last step the generalization capability of the presented neuro-identifier is evaluated by estimating the dynamics of the two-link manipulator by applying a trajectory which is different from the training trajectory.

3.5.1 A Two–Link Manipulator

Simulation results for a two–link planar manipulator are presented in this section. The manipulator [56] consists of one rigid arm (first link) and one flexible arm (second link) with the following numerical data:

$$
\begin{aligned}
l_1 &= 20cm, l_2 = 60cm, A_1 = 5cm \times 0.9mm, A_2 = 3.14cm \times 1.3cm,\\
\rho_1 &= 2700kg/m^3 (6061Aluminum), \rho_2 = 7981(StainlessSteel),\\
M_1 &= 1kg, M_l = 0.251kg, m1 = 0.236kg, m_2 = 0.216kg,\\
E &= 194 \times 10^9 N/m^2, J_1 = 0.11 \times 10^{-3} kgm^2, J_l = 0.11 \times 10^{-4}, J_h = 3.8 \times 10^{-5},
\end{aligned}
$$

where l_1 and l_2 are link lengths, A_1 and A_2 are cross-sectional areas, E and ρ are modulus of elasticity and mass density, J_h is the hub inertia and M_1, M_l, J_1 and J_l are masses and mass moments of inertia at the end points of the two links. The first two natural frequencies of the second link are 5.6 and 27.6Hz

One flexible mode is considered for the simulations. Hence, the state of the system can be defined as $x = [\theta_1\ \theta_2\ \dot{\theta}_1\ \dot{\theta}_2\ \delta_1\ \dot{\delta}_1]^T$ and the input can be defined as $u = [\tau_1\ \tau_2]$ where θ_i and $\dot{\theta}_i$, $i = 1,2$, are the position and velocity of the ith joint; δ_1 and $\dot{\delta}_1$ are the position and velocity of the flexible mode, and τ_i, $i = 1,2$, is the actuator torque of the ith joint. The input to the network is $\hat{\bar{x}} = [\hat{x}\ u]$.

The Hurwitz matrix A is considered as $A = -2I \in \mathscr{R}^{6\times 6}$, the learning rates and damping factors are selected as $\eta_1 = \eta_2 = 1$, $\rho_1 = \rho_2 = 0.01$. A three-layer neural network was used with 8 neurons in the input layer, 10 neurons in the hidden layer, and 6 neurons in the output layer. First, the identifier based on the parallel model was used. The simulation results are shown in Fig. 3.3. Figs. 3.3–a and 3.3–b show the responses of θ_1 and δ_1 to a $sin(t)$ input signal during learning. After the learning has been completed, the responses of $\theta_1, \theta_2, \dot{\theta}_1, \dot{\theta}_2, \delta_1$, and $\dot{\delta}_1$ are shown in Figs. 3.3–c to 3.3–h, respectively. It can be observed that the neural network was able to learn the system dynamics. For all figures, the vertical axis units are rad for joint positions, rad/s for joint velocities, m for the flexible modes, and m/s for the velocities of the flexible modes. Then, the series-parallel identifier was applied. The results are shown in Fig. 3.4. It is clear that significant improvement in terms of the speed of convergence was achieved in this case.

3.5.2 The Space Station Remote Manipulator System (SSRMS)

The Space Station Remote Manipulator System (SSRMS), known as Canadarm 2, is the first component of the Mobile Servicing System which is launched on STS-100 (assembly flight 6A) in April 2001 [87]. It plays a key role in station assembly and maintenance including: moving equipment and supplies around the station, supporting astronauts working in space, and servicing instruments and other payloads attached to the space station. Fig. 3.5 depicts the schematic of

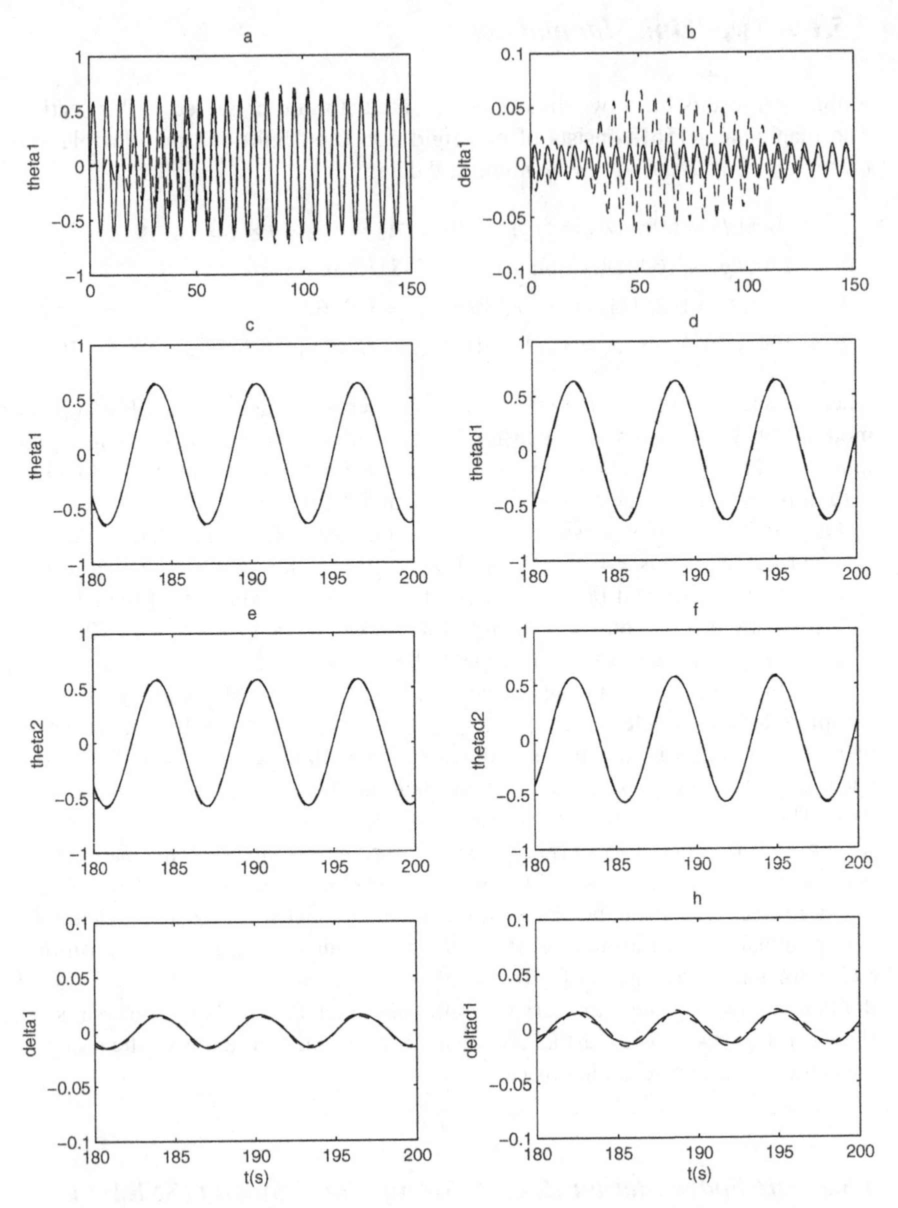

Fig. 3.3 Simulation results for a two-link planar manipulator (parallel model): (a) θ_1 and θ_{n1} during learning, (b) δ_1 and δ_{n1} during learning, (c) θ_1 and θ_{n1} after learning, (d) $\dot{\theta}_1$ and $\dot{\theta}_{n1}$ after learning, (e) θ_2 and θ_{n2} after learning, (f) $\dot{\theta}_2$ and $\dot{\theta}_{n2}$ after learning, (g) δ_1 and δ_{n1} after learning, (h) $\dot{\delta}_1$ and $\dot{\delta}_n$ after learning. The dashed lines correspond to the neural network states.

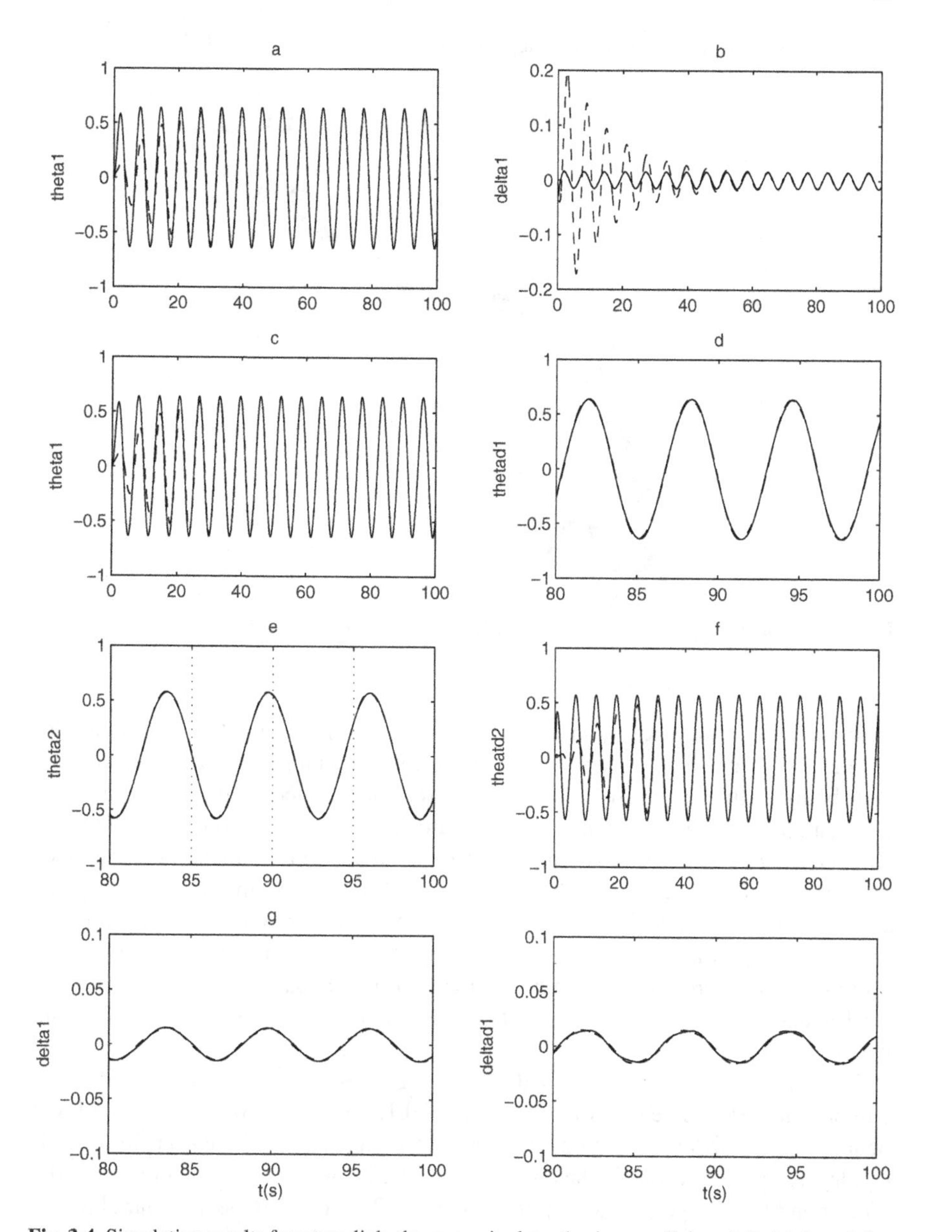

Fig. 3.4 Simulation results for a two-link planar manipulator (series-parallel model): (a) θ_1 and θ_{n1} during learning, (b) δ_1 and δ_{n1} during learning, (c) θ_1 and θ_{n1} after learning, (d) $\dot{\theta}_1$ and $\dot{\theta}_{n1}$ after learning, (e) θ_2 and θ_{n2} after learning, (f) $\dot{\theta}_2$ and $\dot{\theta}_{n2}$ after learning, (g) δ_1 and δ_{n1} after learning, (h) $\dot{\delta}_1$ and $\dot{\delta}_n$ after learning. The dashed lines correspond to the neural network states.

the SSRMS. The SSRMS is a 7 degrees-of-freedom robot which has 7 revolute

Fig. 3.5 Schematic of Canadarm 2 used in the International Space Station [5].

joints and two long flexible links (booms). This is a highly nonlinear system with strong coupling effects. The SSRMS does not have uniform mass and stiffness distributions. In fact, most of its masses are concentrated at the joints, and the joint structural flexibilities contribute a major portion of the overall arm flexibility. For each of the two long links, 1 in-plane, and 1 out-of plane elastic modes were assumed for the model. Hence, the state of the system can be defined as $x = [\theta_1 ... \theta_7 \ \dot{\theta}_1 ... \dot{\theta}_7 \ \delta_{11} \ \delta_{12} \ \delta_{21} \ \delta_{22} \ \dot{\delta}_{11} \ \dot{\delta}_{12} \ \dot{\delta}_{21} \ \dot{\delta}_{22}]^T$ and the input can be defined as $u = [\tau_1, \ ..., \ \tau_7]$ where θ_i and $\dot{\theta}_i$, $i = 1, \ ..., \ 7$ are the position and velocity of the ith joint; δ_{ij} and $\dot{\delta}_{ij}$ are the position and velocity of in plane and out-of plane flexible modes, and τ_i, $i = 1, \ ..., \ 7$ is the actuator torque of the ith joint. The input to the network is $\hat{\bar{x}} = [\hat{x} \ u]$.

The Hurwitz matrix A is defined as $A = -2I \in \mathscr{R}^{22 \times 22}$, the learning rates and damping factors are selected as: $\eta_1 = \eta_2 = 0.1$, $\rho_1 = \rho_2 = 0.001$. A three-layer neural network was used with 29 neurons in the input layer, 20 neurons in the hidden layer, and 22 neurons in the output layer. The 22 outputs of the neural networks correspond to joint positions: 7, joint velocities: 7, in-plane deflection variables: 4, and out-of plane deflection variables: 4. Figs. 3.6 and 3.7 show the simulation results after learning for a $sin(t)$ reference trajectory. For these figures, the dashed lines correspond to the neural network states. As can be observed, despite the high complexity in the manipulator model, the neural network has learned the manipulator dynamics, and all the states of the neural networks track the corresponding states of the system.

3.5.3 Generalization

A common problem found in training neural networks is ensuring that the training patterns are evenly distributed across the problem domain. In the case of training a neural network to identify the dynamics of a flexible-link manipulator, the domain is trajectories. There are several challenges in training neural networks for this problem. One problem is that repeatedly training over a fixed sequence of distributed trajectories results in over-learning the sequence, which has been shown to lead to poor generalization. Considering this issue, another simulation was performed on a two link planar manipulator whose parameters were taken from [56]. The training trajectory for each joint was defined as $0.1(sin(t)+sin(2t)+sin(4t))$. For brevity, only the result for one joint and one flexible mode are given in Figs. 3.8–a to 3.8–d. Later, we stopped the training and applied a totally different trajectory namely $0.15sin(3t)$ to each joint. The results for this case are shown in Figs. 3.8–e and 3.8–h. As can be observed, the proposed neural identifier exhibits a desirable generalization property.

3.6 Experimental Results on a Macro-Micro Manipulator System

Long reach space manipulators such as the Space Station Remote Manipulator System (SSRMS) must be made as lightweight as possible. The reduction in the component mass allows the actuators to move faster and carry heavier loads with longer links and with significantly reduced energy consumption. On the other hand, a lightweight manipulator will have flexibility both in the drive system and in the structure of the manipulator. Structural flexibility causes considerable difficulty in accurate modeling and control of such manipulators. A possible solution is to use a smaller rigid ("micro") manipulator at the end of the larger flexible ("macro") manipulator to achieve the desired accuracy. Such structure is called a *Macro-Micro Manipulator* (M^3) system which was introduced by Sharon and Hardt [88]. Indeed, the small, high-bandwidth rigid manipulator (micro) mounted at the end of a larger link (macro) compensates the inaccuracy of the macro with higher speed. Theses robots has evolved in many practical applications such as tele-robotics [89], nuclear waste cleanup [90, 91], and space robotics [92, 93] which is a combination of SSRMS as macro and Special Purpose Dexterous Manipulator (SPDM) as micro manipulator. These manipulator systems are also called as Flexible Structure mounted Manipulator Systems (FSMS) in the literature [88]. It should be noted that the control of M^3 systems is challenging due to presence of coupling between the two substructures which is regardless of situation of the macro either in motion or stationary situation.

Several researchers have addressed different aspects of M^3 system (e.g. [94, 95, 96, 97, 98]). The experimental set-up consists of a flexible macro and two rigid

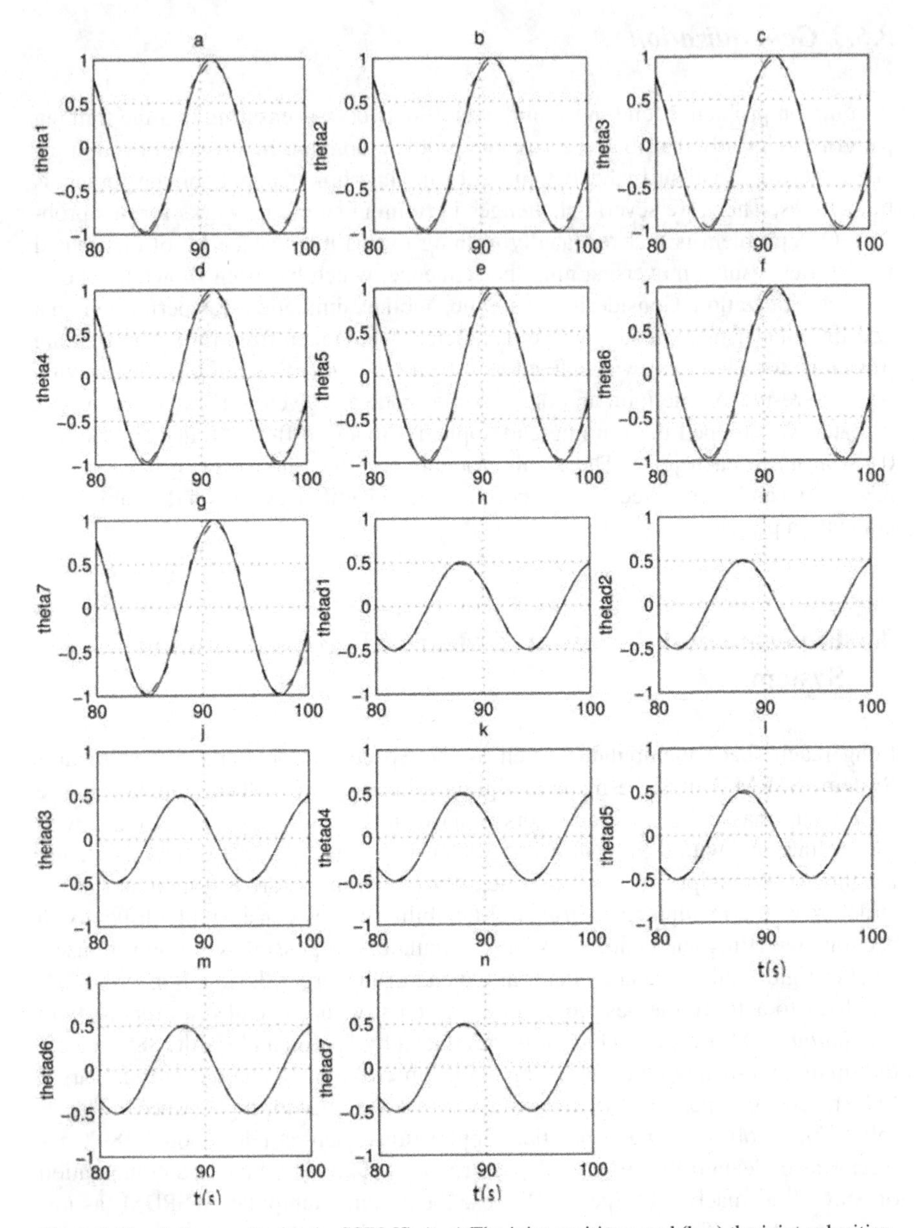

Fig. 3.6 Simulation results for the SSRMS: (a-g) The joint positions, and (h-n) the joint velocities.

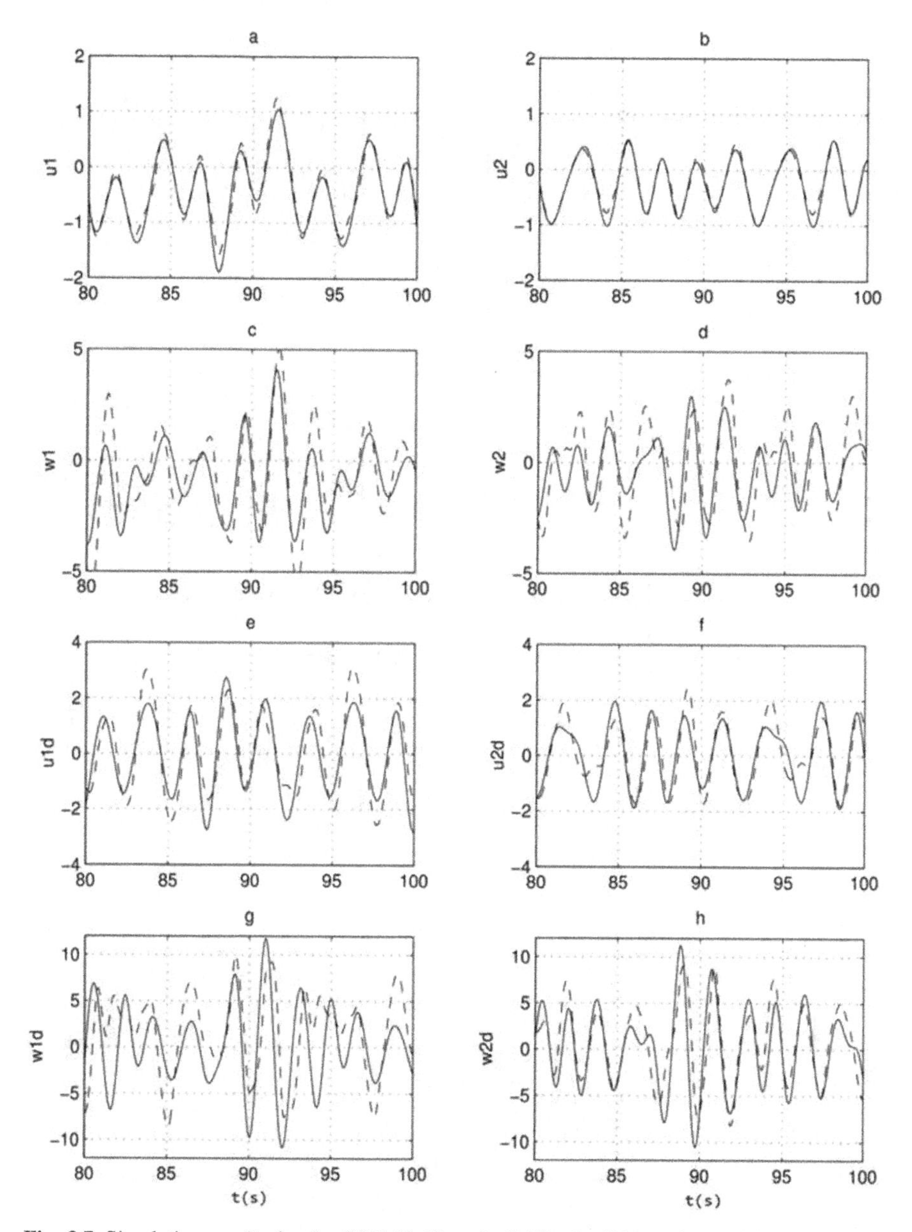

Fig. 3.7 Simulation results for the SSRMS: Figs. (a-d) The flexible modes, and Figs. (e-h) the velocities of the flexible modes.

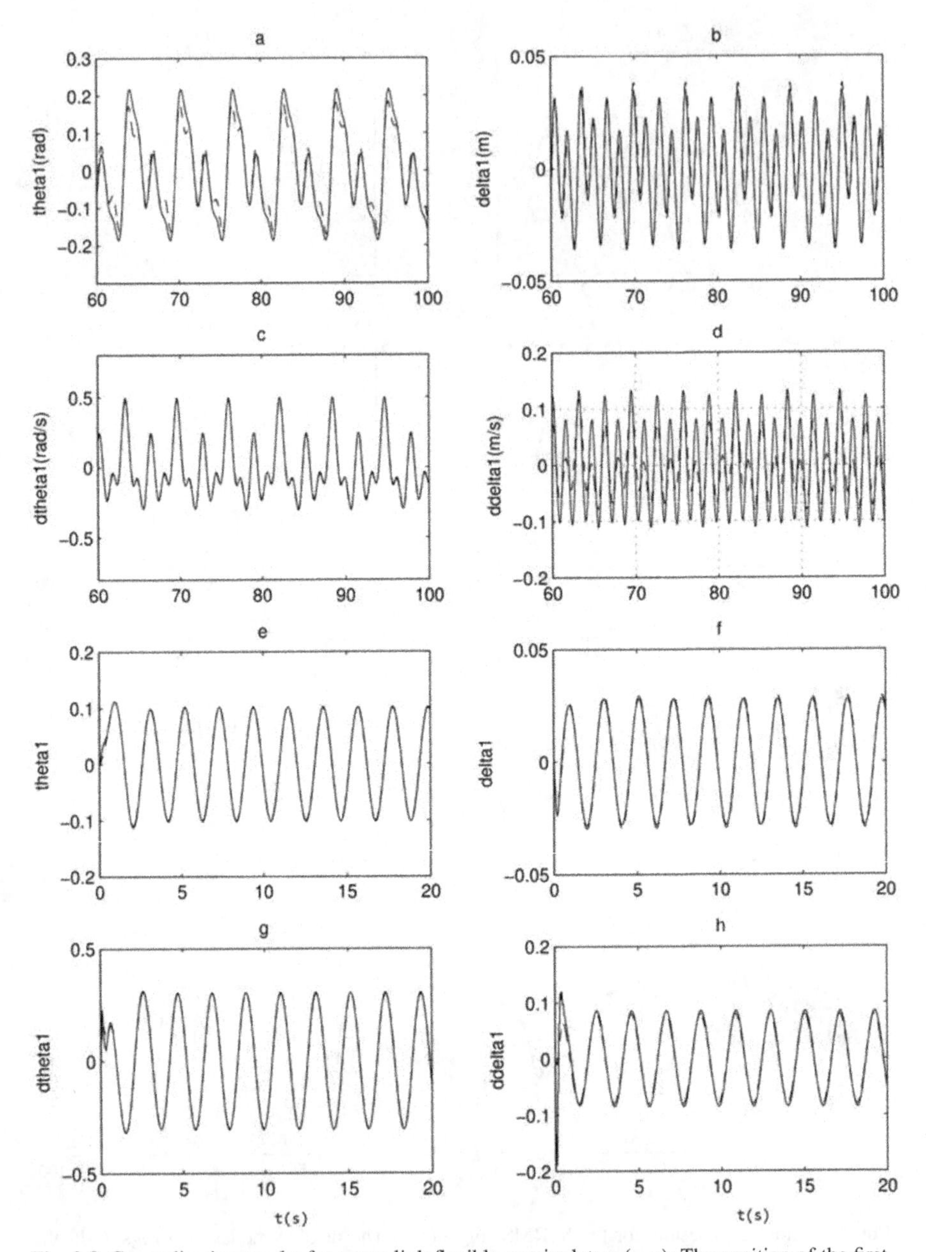

Fig. 3.8 Generalization results for a two-link flexible manipulator: (a, e): The position of the first link; (b, f): The first flexible mode; (c, g): The velocity of the first link; (d, h): The velocity of the first flexible mode. Figs. (a-d) correspond to the training and Figs. (e-h) correspond to the generalization.

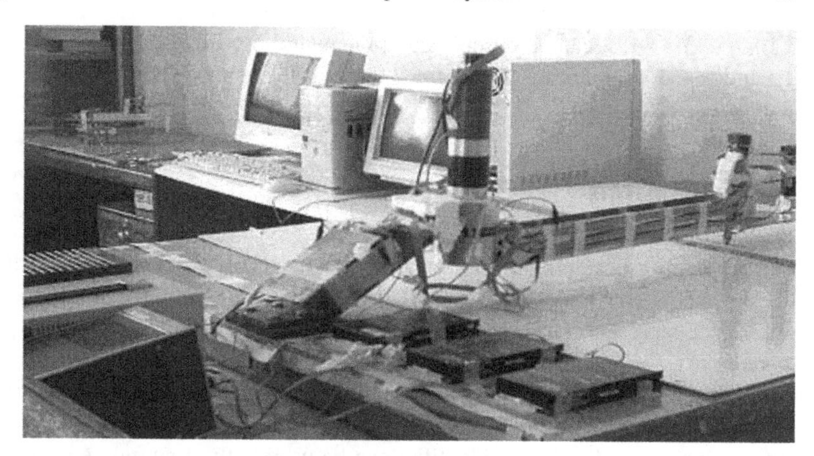

Fig. 3.9 The actual Macro-Micro Manipulator test-bed.

Table 3.1 The parameters of the actual robot

Macro length, width, thickness	$0.59\ m, 70\ mm, 1.25\ mm$
Macro hub radius, inertia	$30\ mm, .63\ Kgm^2$
Macro tip equivalent mass, inertia	$2\ Kg, 0.7\ Kgm^2$
Micro link joint mass, inertia	$0.326\ Kg, 0.003\ Kgm^2$

micro links as shown in Fig. (3.9). The macro link is made of two parallel separated stainless steel beams and the micro links are made of aluminum and have circular cross sections. The actuators are three Maxon DC servomotors equipped with 500 count/rev incremental encoders. All joints have considerable friction. The flexible modes of the Macro link are obtained using strain gauge measurements where only 1 mode is considered for the experiments. The parameters of the test-bed are given in Table 3.1.

In the experiment, we used $A = -3I \in \mathscr{R}^{8\times 8}$, $\eta_1 = \eta_2 = 1$, $\rho_1 = \rho_2 = 0.001$. The proposed neural network identifier is implemented in Matlab XPC-Target environment using the Watcom C compiler. A PCL818H data acquisition card from ADVANTEC was used with a sampling period of $1ms$. The neural network used for the experiments is a three layer network with 11 neurons in the input layer, 10 neurons in the hidden layer and 8 neurons in the output layer. The inputs to the network are $\theta, \dot{\theta}, \delta, \dot{\delta}$, and u. The hidden layer neurons have tangent hypebolic transfer function and the output neurons use the linear activation function.

For the experiments, the macro joint was locked and the two micro joints were moved with $sin(t)$ and $cos(t)$ trajectories, respectively. The proposed identification scheme was implemented and the results are shown in Fig. 3.10. Figs. 3.10–a, 3.10–c and 3.10–e show the responses of θ_2, θ_3 and δ_1 respectively, where θ_2 and θ_3 are for the micro joints and δ_1 is the first flexible mode of the macro link. The velocities of the joints of the manipulator and that of the flexible mode are also shown in Figs.

3.10–b to 3.10–f. It can be observed that the neural network has learned the system dynamics and its states accurately track those of the actual robot.

3.7 Conclusions

The problem of designing a stable identification scheme for a general MIMO nonlinear system using both LPNN and NLPNN has been investigated in this chapter. A state space representation was considered based on both parallel and series-parallel models. The nonlinear-in-parameters characteristic of the proposed identifier makes it a powerful tool in identifying the nonlinear system dynamics with arbitrary degrees of complexity and nonlinearity and with no *a priori* knowledge about their system dynamics. The neuro-identifier can be used both as an online as well as an off-line scheme. Finally, the proposed identifier was implemented on an experimental test-bed consisting of a complicated nonlinear system, namely a Macro-Micro Manipulator system. Experimental results confirm the reliability of the performance of the stable identification scheme.

The next three chapters present another application of the proposed neural network observer/identifier in fault detection and isolation.

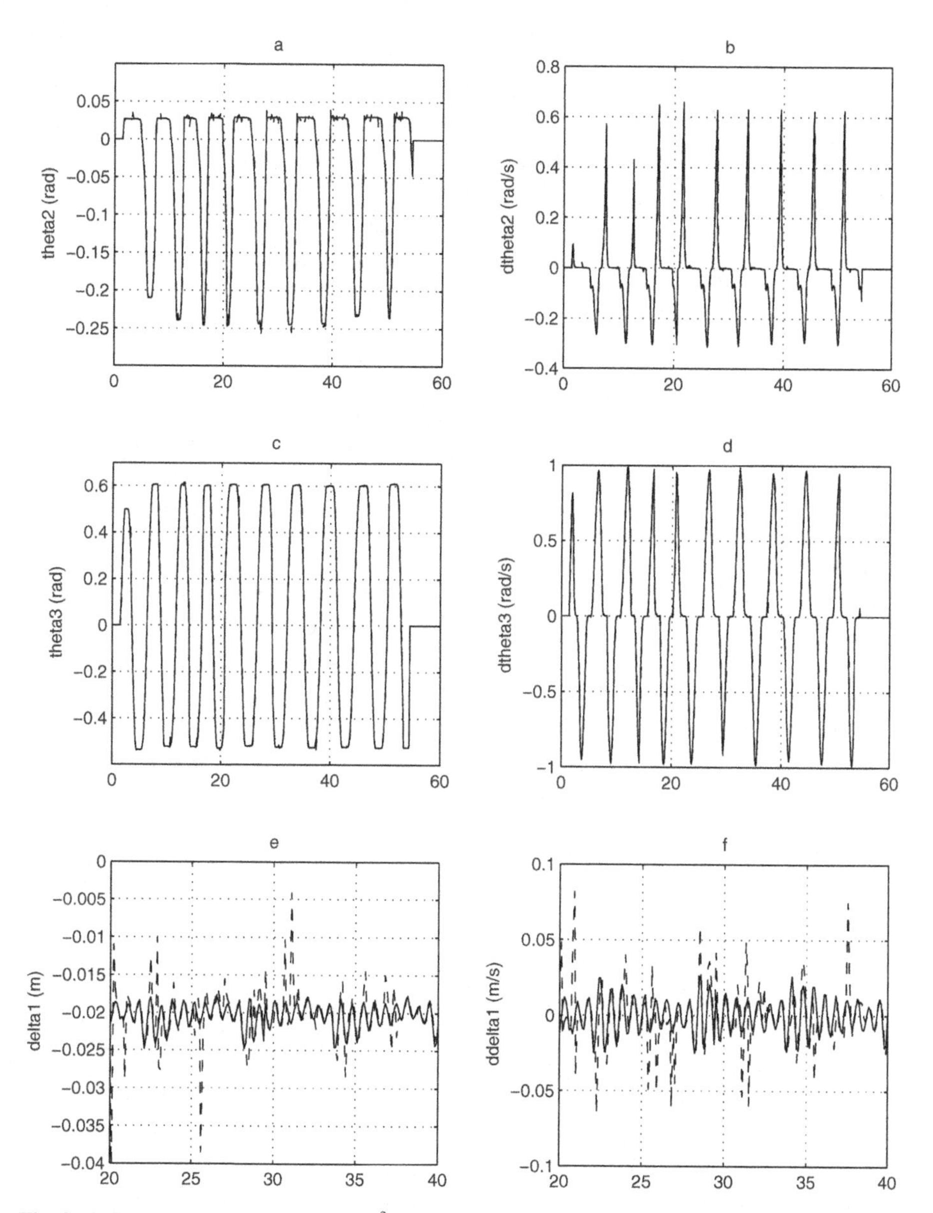

Fig. 3.10 Experimental results for the M^3 test-bed using the proposed neuro identifier: (a-b) Joint position and velocity of the first micro link, (c-d) Joint position and velocity of the second micro link and (e-f) First flexible mode of the macro link and its velocity. The dashed lines correspond to the neural network states.

Chapter 4
An Actuator Fault Detection and Isolation Scheme: Experiments in Robotic Manipulators

4.1 Introduction

The problem of fault detection, isolation and recovery (FDIR) has received a great deal of attention during the past several years. There are a number of industrial applications such as manufacturing processes, hazardous waste management, and maintenance of aircraft and electrical power lines where unpredictable behavior such as poor performance or even unsafe operation can result from abnormal deviations of components or subsystems. In certain industrial applications unpredictable behaviors are usually associated with increasing operating costs, line shut-down, and possible detrimental environment impacts. Consequently, an exceptional level of autonomy is required in these types of applications. Hence, there is a high demand for new methods in supervision, fault detection, fault isolation, and fault recovery. Fault detection and identification is an essential component of an autonomous system. Each FDI procedure consists of three consequent stages: (i) detection of the fault occurrence, (ii) fault diagnosis in which the type, location and magnitude of detected fault are determined, and (iii) fault accommodation in which fault impacts are restricted or eliminated in a proper manner based on decisions made in second stage.

There are a number of approaches proposed for automated fault diagnosis in nonlinear systems. For instance, expert system methodologies [99] and time series analysis [100] among others are used for failure detection purposes. However, the two main approaches that use analytical redundancy are state and parameter estimation techniques. Parameter estimation methods require knowledge of the model structure of the investigated process and actual process measurements. The actual values of some critical model parameters (e.g. the mass matrix in a robotic system) are then determined from this information. The fault detection is accomplished by a comparison of identified and predetermined values by using some thresholds [101]. Observer-based methods require knowledge of the investigated process as well. The model operates parallel to the real process. Assuming a nominal model of the plant, the difference between measured and calculated desired variables (e.g.

H.A. Talebi et al., *Neural Network-Based State Estimation of Nonlinear Systems*,
Lecture Notes in Control and Information Sciences 395,
DOI 10.1007/978-1-4419-1438-5_4,

joint velocities in robotic systems) will produce a non-zero value when a fault has occurred. This difference is called a residual. A fault is declared if the residual exceeds a certain threshold value. The main advantage of observer-based methods over parameter estimation methods results from the fact that no special excitation (persistent excitation) is required. The observer-based methods also work in steady state operating conditions. It is also well known that the fault capability of the nonlinear observer-based fault detection approach is significantly better than that of the linear observer-based approach. Moreover, dynamic faults often cause unpredictable nonlinear changes in the dynamics of the system. Therefore, to capture a large class of practical failure situations, a nonlinear modeling framework is required.

Fault detection and isolation deal with estimation and classification problems. The capability of neural networks for function approximation and classification and their ability to deal with uncertainty and/or parameter variations make them viable choices for the fault detection problem. Hence, neural networks can be employed in the state/parameter estimation and/or fault isolation step. For instance, the function approximation ability of neural networks was used in [102] to solve the kernel-coupled nonlinear equation required for machine fault detection based on Higher Order Statistic (HOS) analysis. Using a neural network in the estimation step [103, 104] requires the network to identify a dynamical system. If an off-line training scheme is used, the main issue is to provide the network with proper training sets. This often requires that the reference trajectories are known *a priori*. Using an online training scheme, on the other hand, adds extra complexity to the problem. The main difficulty here is distinguishing between changes in the normal working condition and the faulty behavior of a system such as a robot.

In [105], a neural network classifier is trained off-line to identify the type of faults from the residual generated by a nonlinear observer which assumes an exact knowledge of mathematical model of the system. No stability analysis was given. However, simulation results for a robot manipulator were provided. In [16], fault detection and estimation approach was developed for robot manipulators. A nonlinear observer was employed based on a mathematical model of the robot . The fault function was estimated by utilizing a Linear-in-Parameter Neural Network (LPNN), the weights of which were updated by means of a projection operator and only simulation results were reported. In [106], a fuzzy logic approach is employed to detect the faults by using the residual generated from a model-based nonlinear observer which lacks a stability analysis. In [107], an Radial Basis Function (RBF) neural network was used to identify and classify the faults in an induction machine based on four feature vectors extracted from power spectra of the machine vibration signals. A fault diagnosis scheme for power generation plants was introduced in [108] by using a neuro-fuzzy approach.

Sensor and actuator fault detection and identification for nonlinear systems were considered in [17, 109, 110, 18, 111, 19, 112, 20]. A supervisory switching logic algorithm was used in [113] to reconfigure the attitude control system in the case of actuator failures. Depending on the state of the actuator, an appropriate controller was selected from a bank of controllers. In [114, 115, 116, 117], several neural network-based fault detection schemes for satellite attitude control subsystems were developed. The weights of neural networks were updated using the Extended Kalman Filter (EKF) algorithm. However, stability analysis relied on the convergence of the EKF which in turn requires certain restrictive assumptions to hold.

In this chapter, the neural network-based identification and observation approaches proposed in the previous chapters are employed for designing an actuator fault detection and identification scheme for nonlinear systems. The fault detection scheme is based on a hybrid model composed of an analytical model and an intelligent model of the nonlinear system. The nominal performance of the system in fault-free operation is governed by analytical model. Whereas, the uncertainties and unmodeled dynamics are accounted by intelligent (neural network-based) model. Another nonlinear-in-parameters neural network (NLPNN) is employed to identify an unknown fault. The neural network weights are updated based on modified backpropagation scheme. The stability of the overall fault detection scheme is shown using the Lyapunov's direct method. The effectiveness of the proposed fault diagnosis strategy is first demonstrated via simulations carried out on a highly accurate dynamical model of the reaction wheel actuator. The performance of the FDI scheme is also experimentally verified on an industrial robot, namely the PUMA 560 manipulator. The main advantage of the proposed FDI strategy is the weight-updating mechanism which benefits from easy implementation of the backpropagation plus increased robustness obtained by adding the e-modification term. Moreover, fault detection, isolation, and estimation steps are all unified, i.e., neither extra measured/calculated signals nor a separate fault isolation policy is required to isolate the faults.

The rest of the chapter is organized as follows: Section 4.2 presents the neural network structure for fault detection and isolation. Two case studies are considered to show the performance of the proposed fault detection scheme. First, in Section 4.3 a satellite Attitude Control System (ACS) with reaction wheel type of actuators is considered and simulation results are given. Next, the fault detection problem in robotic manipulators is considered and experimental results are presented in Section 4.4. Finally, Section 4.5 concludes the chapter.

4.2 Neural Network Structure for Actuator Fault Detection

Consider the nonlinear system

$$\dot{x} = f(x,u) + N_1(x,u) + T_F(x,u,t), \tag{4.1}$$

where $u \in R^m$ is the input, $x \in R^n$ is the state vector of the system. Note that in the nonlinear system represented by a hybrid model (4.1), f is a known nonlinear function representing the nominal model, N_1 is the mapping performed by the first neural network NN_1 (see Fig. 4.1), and T_F is a vector-valued function of an unknown actuator fault. NN_1 is trained in fault-free operation mode to compensate for the nominal model deficiencies arising from unmodeled dynamics and uncertainties.

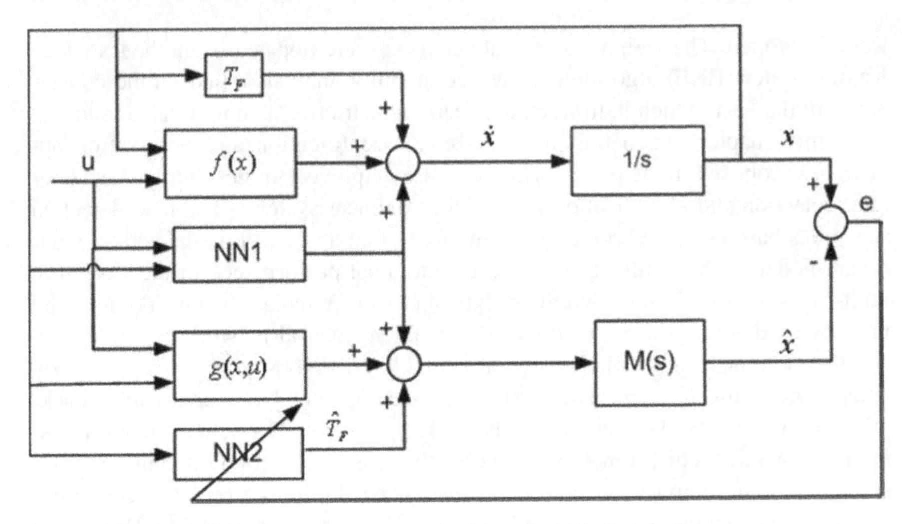

Fig. 4.1 Structure of the neural network estimator for actuator fault detection.

Now, by adding and subtracting Ax, we can write (4.1) as

$$\dot{x} = Ax + g(x,u) + N_1(x,u) + T_F(x,u,t), \tag{4.2}$$

where $g(x,u) = f(x,u) - Ax$ and A is a Hurwitz matrix. In the light of Theorem 1.1, the fault function T_F can be expressed as

$$T_F(x,u,t) = W\sigma(V\bar{x}) + \varepsilon(x), \tag{4.3}$$

where W and V are the fixed ideal weight matrices, $\bar{x} = [x \quad u]$, and $\varepsilon(x) \leq \varepsilon_N$ is the neural network's bounded approximation error. The function $T_F(x,W,V)$ can be approximated as

$$\hat{T}_F(x,\hat{W},\hat{V},t) = \hat{W}\sigma(\hat{V}\bar{x}). \tag{4.4}$$

According to the representation (4.2), a recurrent network model is now constructed by parameterizing the mapping T_F through utilizing feedforward (static) neural network architectures, denoted by NN_2 (see Fig. 4.1). Specifically, the following model is considered for our observer-based design strategy:

$$\dot{\hat{x}} = A\hat{x} + g(x,u) + N_1(x,u) + \hat{T}_F(\bar{x},\hat{W},\hat{V},t), \tag{4.5}$$

where $\hat{T}_F(\bar{x},\hat{W},\hat{V})$ is the mapping performed by the neural network NN_2 as given in (4.4). The structure of the estimator is shown in Fig. 4.1. Corresponding to the Hurwitz matrix A, $M(s) := (sI - A)^{-1}$ is also shown which is an $n \times n$ matrix whose elements are stable transfer functions. Similar to [16], the neural network here is employed to identify the fault and produce zero output in non-faulty operation. When there is a fault, the output of the neural network $\hat{T}_F$ is nonzero and that is used as a residual to identify the type of fault.

Substituting (4.4) in (4.5) results in the estimator model:

$$\dot{\hat{x}}(t) = A\hat{x} + g(x,u) + N_1(x,u) + \hat{W}\sigma(\hat{V}\bar{x}). \tag{4.6}$$

By defining the estimation error as $\tilde{x} = x - \hat{x}$, and using (4.2), (4.4) and (4.6), the error dynamics can be stated as

$$\dot{\tilde{x}}(t) = A\tilde{x} + \tilde{W}\sigma(\hat{V}\bar{x}) + w(t), \tag{4.7}$$

where $\tilde{W} = W - \hat{W}$ is the neural network's output weight error, $w(t) = W[\sigma(Vx) - \sigma(\hat{V}x)] + \varepsilon(x)$ is a bounded disturbance term, i.e., $\|w(t)\| \leq \bar{w}$ for some positive constant $\bar{w}$, due to the tangent hyperbolic function.

Indeed, by employing the series-parallel neuro-identifier introduced in Section 3.3, the fault function T_F can be identified and eventually a stable fault detection scheme be provided. The following theorem summarize the results.

Theorem 4.1. *Consider the nonlinear system (4.1) and the identifier model (4.5). If the weights of NLPNN are updated according to:*

$$\dot{\hat{W}} = -\eta_1(\frac{\partial J}{\partial \hat{W}}) - \rho_1\|\tilde{x}\|\hat{W} \tag{4.8}$$

$$\dot{\hat{V}} = -\eta_2(\frac{\partial J}{\partial \hat{V}}) - \rho_2\|\tilde{x}\|\hat{V}, \tag{4.9}$$

then $\tilde{x}, \tilde{W}$, and $\tilde{V} \in L_\infty$, i.e., the estimation error and the weights error are ultimately bounded. In these equations, $\eta_1, \eta_2 > 0$ are the learning rates, ρ_1, ρ_2 are some small positive numbers, and $J = \frac{1}{2}(\tilde{x}^T\tilde{x})$ is the neural network objective function.

Proof: The proof follows along the similar lines to the proof of Theorem 3.2.

Remark 4.1 *The proof of the theorem completes the development of the proposed methodology that can detect, isolate, and estimate the fault vectors T_F in a stable fashion with* no *restrictive assumption. Note that if the fault vectors are estimated correctly, the isolation and detection objective are automatically satisfied. One of the main advantages of the proposed scheme is that the fault detection, isolation, and estimation steps are all unified. That is to say, the signals generated by the neural networks serve as residuals as well as estimated faults in the corresponding actuator. Hence, neither extra measured/calculated signals nor a separate fault isolation policy is required to isolate the fault.*

Remark 4.2 *It is apparent from (4.5) that the accuracy of the FDI scheme is directly related to that of neural network estimation, i.e., $\hat{T}_F$ which is in turn determined by the accuracy of state and weight estimation. There are different parameters that can be adjusted to enhance the accuracy of the neural network estimation e.g. the learning rates η_1 and η_2, the damping factors ρ_1 and ρ_2, and the Hurwitz matrix A.*

4.3 Case Study 1: Application to Satellite Attitude Control Subsystems

The improvements in the accuracy and reliability of Attitude Control Systems (ACS) contributes directly to the success and reliability of satellites in space. Hence, an exceptional level of autonomy is required. The inherent nonlinearity of ACS dynamics, however, makes the accurate and efficient fault detection of ACS a challenging problem. In this section, the proposed fault detection and isolation scheme is employed to detect actuator fault in a satellite attitude control system.

4.3.1 System Dynamics

The dynamic equations of a satellite is described by

$$H\dot{\omega} = -S(\omega)H\omega + \tau, \tag{4.10}$$

where $\tau = [\tau_x \ \tau_y \ \tau_z]^T$ is the torque vector provided by three reaction wheels in x, y, and z channel respectively, $\omega = [\omega_1 \ \omega_2 \ \omega_3]^T$ is the angular velocity vector of the satellite, expressed in satellite body frame, H is the symmetric positive definite inertia matrix of the satellite and $S(\omega)$ is the cross product matrix given by

$$S(\omega) = \begin{bmatrix} 0 & \omega_3 & -\omega_2 \\ -\omega_3 & 0 & \omega_1 \\ \omega_2 & -\omega_1 & 0 \end{bmatrix}. \tag{4.11}$$

The attitude of the spacecraft can be represented in many ways using ,e.g., Euler angles or singularity-free Euler parameters leading [118] to

$$\Omega = \Pi(\Omega)\omega,$$

where the definitions of Ω and Π depend on the choice of the kinematic representation. Although, the fault diagnosis algorithm proposed in this chapter is independent of the choice of attitude representation, the Euler angle representation is considered for simulation purposes. Hence, the kinematic equations are described by

$$\begin{bmatrix} \dot{\Psi} \\ \dot{\theta} \\ \dot{\Phi} \end{bmatrix} = \Pi(\Psi,\theta,\Phi) \begin{bmatrix} \omega_1 \\ \omega_2 \\ \omega_3 \end{bmatrix} \tag{4.12}$$

$$\Pi(\Psi,\theta,\Phi) = \begin{bmatrix} 0 \ S\Phi/C\theta & C\Phi/C\theta \\ 0 \quad C\Phi & -S\Phi \\ 1 \ S\Phi tan\theta & C\Phi tan\theta \end{bmatrix},$$

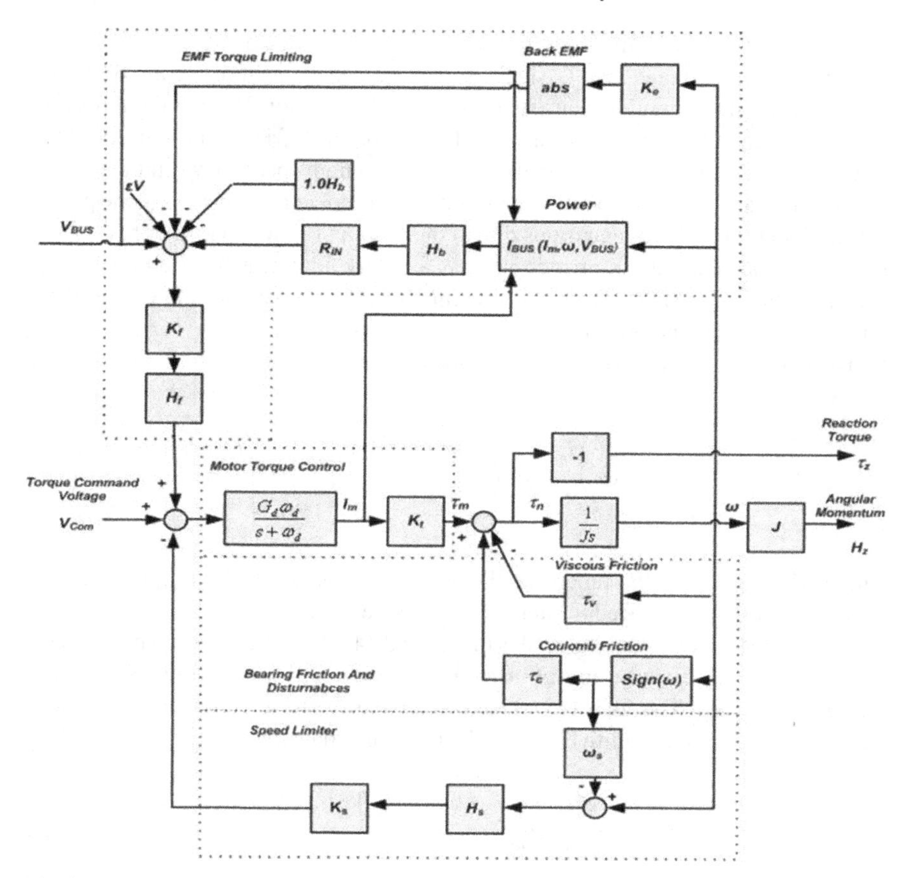

Fig. 4.2 Schematic diagram of the reaction wheel.

where $\Omega = [\Psi \ \theta \ \Phi]^T$ is the Euler angle vector representing the orientation of the satellite body frame and C, S and tan denote the cos, sin and $tangent$ functions, respectively.

4.3.2 Reaction Wheel Model

Reaction/momentum wheels are flywheels used to provide attitude control authority and stability on satellites. By adding or removing energy from the flywheel, torque is applied to a single axis of the satellite, causing it to react by rotating. A reaction wheel is basically just a spinning flywheel mounted on a central bearing whose rate of rotation can be adjusted as necessary by an electric motor to apply a force and move the satellite. Fig. 4.2 shows the mathematical model of the reaction wheel in block diagram form which is taken from [18]. In this figure, K_t is the motor torque constant, I_m is the motor current, ω_m is the motor angular velocity, V_{com} is the torque command voltage, and G_d is the transconductance gain of the motor driver. The speed limiting circuitry consists of a threshold function H_s with the threshold ω_s and the gain K_s. The model also includes Coulomb friction (τ_c) and viscous friction (τ_v). The increase in back-EMF effect, K_e at high speed may limit the motor torque especially at low bus voltage conditions. This effect also coupled to the power consumption through voltage drops due to the product of the input filter resistance, R_{in} and the bus current I_{BUS} which itself is a highly nonlinear function of I_m, ω_m, and bus voltage V_{BUS} [18]. The heavyside function H_b compensates the voltage drop for situations where no power is drawn from the bus.

The mathematical description of the block diagram model shown in Fig. 4.2 can be given by:

$$\begin{aligned}\begin{bmatrix} \dot{I}_m \\ \dot{\omega}_m \end{bmatrix} &= \begin{bmatrix} G_d\omega_d\left[\Psi_1(I_m,\omega_m) - \Psi_3(\omega_m)\right] - \omega_d I_m \\ \frac{1}{J}\left[K_t I_m - \tau_c\Psi_2(\omega_m) - \tau_v\omega_m\right] \end{bmatrix} + \begin{bmatrix} G_d\omega_d \\ 0 \end{bmatrix} V_{com} \\ \tau &= K_t I_m,\end{aligned} \tag{4.13}$$

where Ψ_1, Ψ_2, and Ψ_3 represent the nonlinearities for EMF torque limiting, Coulomb friction, and speed limiter subsystems, respectively.

Now, consider the satellite model (4.10) and (4.12) and reaction wheel model (4.13). For this case study, an **unknown fault** in reaction wheel command voltage V_{com} is considered. The fault is included as an extra voltage component V_F which basically appears as an additive term to the reaction wheel dynamic equation (4.13):

$$\begin{aligned}\dot{I}_m &= G_d\omega_d\left[\Psi_1(I_m,\omega_m) - \Psi_3(\omega_m)\right] - \omega_d I_m \\ &\quad + G_d\omega_d(V_{com} + V_F),\end{aligned} \tag{4.14}$$

where V_F represents the unknown fault. Now, by defining $x = \begin{bmatrix} \Omega \\ \omega \\ I_m \\ \omega_m \end{bmatrix}$ and $u = V_{com}$,

Equations (4.10), (4.12), (4.13), and (4.14) can be written as

$$\begin{aligned}\dot{x} &= f(x,u) + T_F \\ T_F &= \begin{bmatrix} 0 \\ 0 \\ G_d\omega_d V_F \\ 0 \end{bmatrix}.\end{aligned} \tag{4.15}$$

It can be seen that (4.15) is in the form of (4.1), hence the methodology developed in Section 4.2 can be applied to detect the fault vector T_F.

4.3.3 Simulation Results

To verify the performance of the proposed fault diagnosis scheme, simulations are performed and the results are presented in this section. The inertia matrix and the reaction wheel parameters (Table 4.1) used in the simulations were adopted from [119] and [111], respectively:

$$H = \begin{bmatrix} 5.5384 & -.0276 & -.0242 \\ -.0276 & 5.6001 & -.0244 \\ -.0242 & -.0244 & 4.2382 \end{bmatrix} kg.m^2. \tag{4.16}$$

A conventional state feedback controller is used to stabilize the closed-loop system

Table 4.1 Reaction wheel parameters

Parameter	Value
J	$0.0077\ N.m.s^2$
G_d	$0.19\ A/V$
K_t	$0.029\ N.m/A$
K_e	$0.029\ V/rad/s$
K_s	$95\ V/rad/s$
ω_s	$690\ rad/s$
τ_c	$0.002\ N.m$
N	36
ω_d	$20\ rad/s$
R_m	$2\ \Omega$
K_f	$0.5\ V/V$

in fault-free operation. The neural network has three layers, the input layer has 12 neurons, the hidden layer has 10 neurons with tangent hyperbolic activation functions, and the output layer has 3 neurons with linear transfer functions. The input of the network is x. The output of the network is $\hat{V}_F = [\hat{V}_{Fx}\ \hat{V}_{Fy}\ \hat{V}_{Fz}\ ]^T$. The numerical values used for the neural networks learning parameters in (4.8) and (4.9) are: $\eta_1 = \eta_2 = .1; \rho_1 = \rho_2 = 10^{-6}$, and the Hurwitz matrix A is selected as $A = -5I_{12}$, where I_{12} denotes a 12×12 identity matrix.

At first, a 50% uncertainty in motor torque constant K_t is incorporated and all sensory measurements are corrupted with 10% Gaussian noise. A $0.1sin(0.01t)$ current reference trajectory was applied to the reaction wheel in fault-free operation and the responses of the current as well as the angular velocities of the wheel are shown in Fig. 4.3-a to 4.3-f, respectively. It can be seen that there is a considerable mismatch between the actual and estimated angular velocities due to the presence of uncertainty in motor torque constant K_t. To resolve this issue, a neural network, namely N_1 as introduced in Section 4.2 was employed to identify the unmodeled dynamics of the reaction wheel in its normal operating condition. Next, the reaction wheel was commanded with the same reference trajectory in fault-free situation and the responses of the wheel angular velocities are depicted in Fig. 4.4-a to 4.4-c. It is demonstrated that, a good match is obtained by adding N_1 to the estimator dynamics. The output of the neural network estimating the fault, i.e., $\hat{V}_F$ are also shown in Fig. 4.4-d to 4.4-f. As can be observed, the output of the network goes to a small value very close to zero confirming that there is no fault in the system.

Next, an actuator fault as a constant voltage $(1V)$ in each channel is introduced to the system. The responses of the fault detection system are shown in Fig. 4.5. It can be observed that the output of the neural network approaches to a nonzero value indicating that there is a fault in the system. In fact, the magnitude of the torque is exactly the same as that introduced in the system. The angular velocities are also shown in Fig. 4.5 which indicates that once the fault is estimated, the states of the observer follow the states of the system very accurately. In the next simulation, we attempted to evaluate the sensitivity as well as separability of the fault detection scheme. Hence, we introduced a constant fault with very *small* magnitude, i.e., $(V_F = .01$ V$)$ *only* in channel x actuator. The response of the neural network output as well as estimation errors are shown in Fig. 4.6. The figure shows that although the fault detection system was able to correctly estimate the fault in channel x, it takes a longer time for the neural network to converge due to the fact that the fault signature on angular velocity was so small. One remedy to speed up the convergence is to increase the learning rates η_1 and η_2 with caution since larger values of learning rates tend to destabilize the system especially for faults with larger magnitudes. In the last simulation, transient behavior of the fault detection system subject to temporary faults was considered. Hence, a 1 N-m fault was introduced in channel x actuator from $t = 200$ *second* to $t = 800$ *second*. The output of the neural network, and angular velocities are shown in Fig. 4.7. These figures confirm that the neural network was able to recognize the existence of the fault inserted in the system at $t = 200$ *second*. Moreover, its output converged to zero around $t = 800$ *second* indicating that the fault has been removed from the system.

Next, a fault in the bus voltage is introduced in the system, namely the bus voltage is changed from its nominal value of 24 V to 9 V in each channel. The responses of the fault detection system are shown in Figure 4.8. It can be observed that the output of the neural network approaches to a nonzero value showing that there is a fault in the system. The motor currents are also shown in Figure 4.8-a to 4.8-f which indicates that the states of the neural network follow the states of the system very accurately. However, in this case the control is not reconfigured and it can be seen that there is a significant of degradation in the responses of the motor currents due to the drop in the bus voltage. In the last simulation, the output signal of the neural network (the estimated fault) was used to reconfigure the controller. In fact, the controller is now given by $V_{com} - \hat{V}_F$. The motor currents and the estimated faults are shown in (Figures 4.9-a to 4.9-f). It can be observed that considerable improvement is obtained when the control reconfiguration is used as compared to 4.8-a to 4.8-f. These experiments confirm that the neural network was able to recognize the existence of the fault and successfully reconfigure the controller to compensate for the fault inserted in the system.

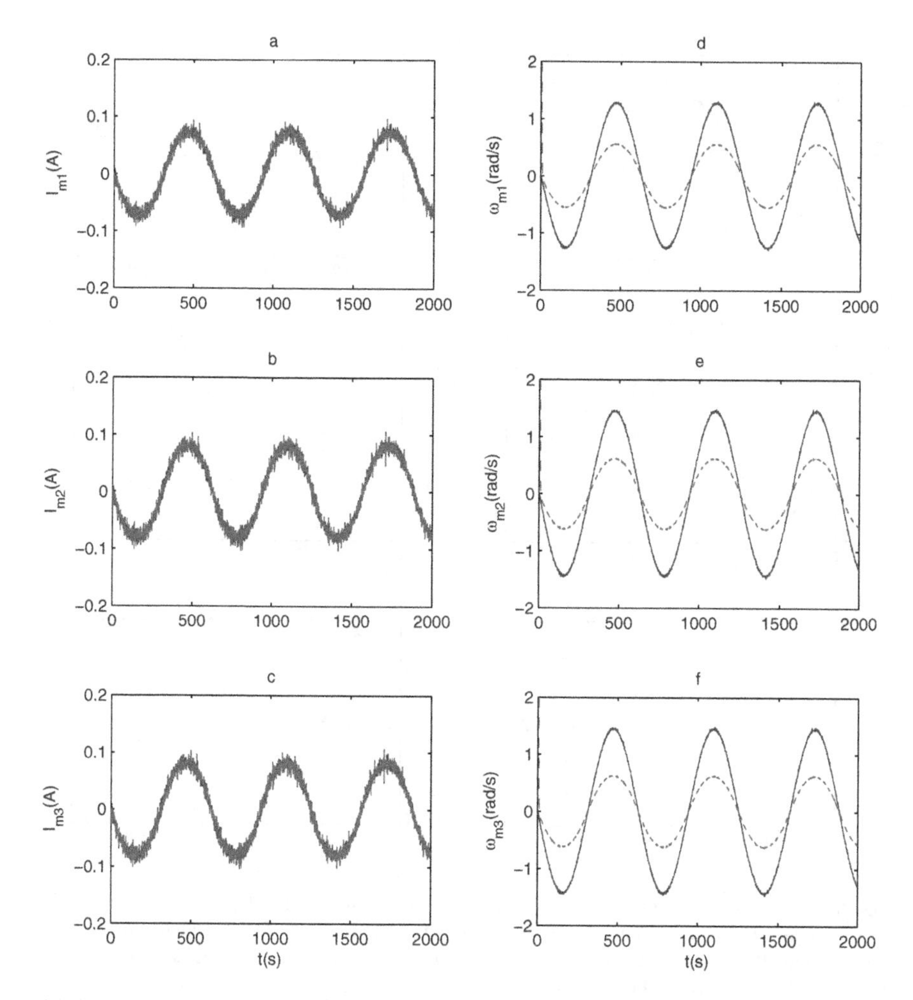

Fig. 4.3 The responses of current and angular velocities of the wheel to $0.1sin(0.01t)$ reference trajectory in fault-free operation in the presence of 50% uncertainty in K_t: (a)-(c) actual and estimated current; (d)-(f) actual and estimated angular velocities. The solid lines correspond to the actual states and the dashed lines correspond to their estimates.

4.4 Case Study 2: Application to Robotic Manipulators

There are a number of industrial applications such as manufacturing processes, hazardous waste management, and maintenance of aircrafts and electrical power lines where robotic manipulators require to operate in remote and hazardous environments. In this section, the proposed methodology is applied to a fault detection and estimation problem in robotic manipulator systems.

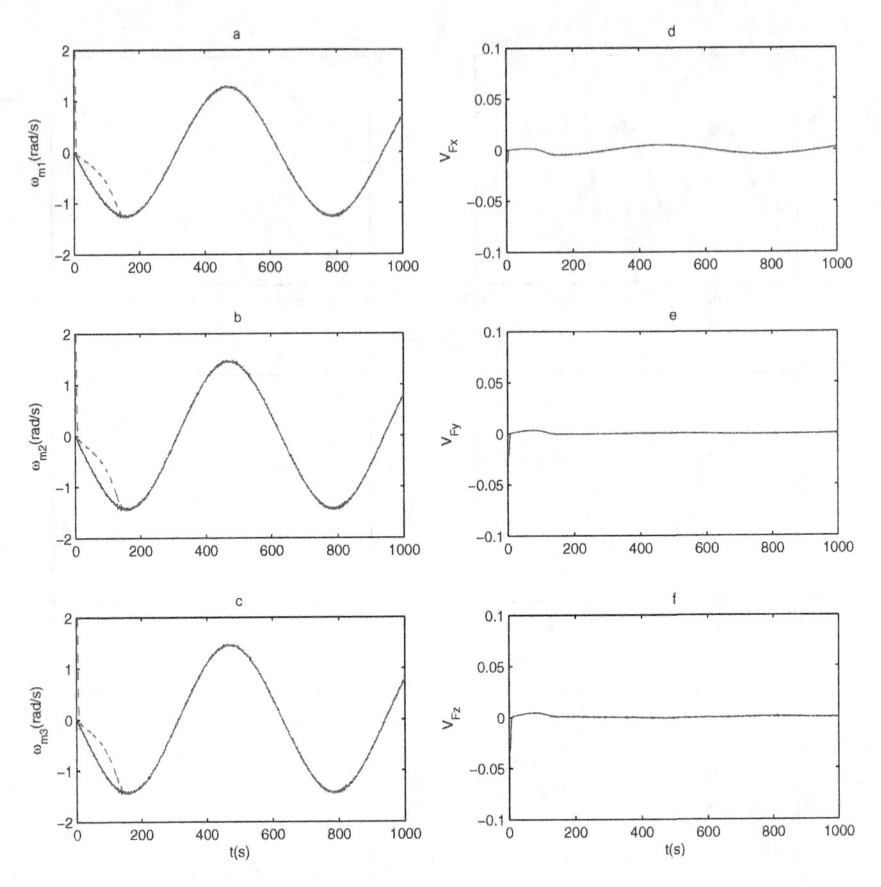

Fig. 4.4 The responses of the reaction wheel and the hybrid model to $0.1sin(0.01t)$ reference trajectory in fault-free operation in the presence of 50% uncertainty in K_t: (a)-(c) actual and estimated angular velocities; (d)-(f) estimated faults. The solid lines correspond to the actual states and the dashed lines correspond to their estimates.

4.4.1 System Dynamics

The dynamic equations of a robot manipulator is given by

$$M(q)\ddot{q}+C(q,\dot{q})+G(q)=\tau+\tau_F, \tag{4.17}$$

where $q \in R^n$ is the manipulator's joints position variable, $\dot{q} \in R^n$ is the joint velocity, τ is the applied torque, and τ_F represents an unknown fault. Now, by defining $x = [q^T, \dot{q}^T]^T$, Equation (4.17) can be written as

$$\dot{x} = f(x,u)+T_F, \tag{4.18}$$

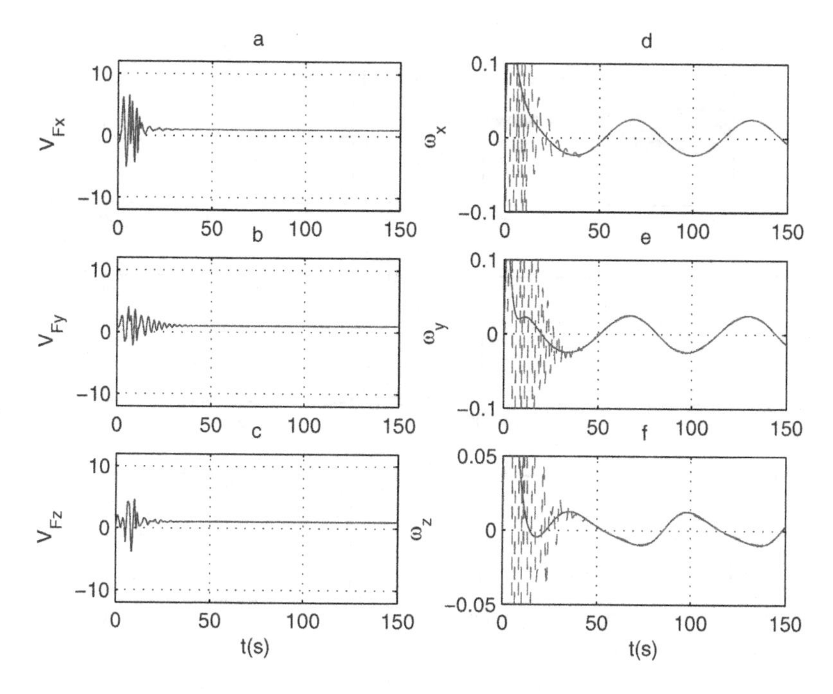

Fig. 4.5 Angular velocities of the satellite to $0.2sin(0.1t)$ reference trajectory with $1.0N.m$ constant fault in all actuators: (a)-(c) estimated faults; (d)-(f) satellite angular velocities, ω_x, ω_y, ω_z. The solid lines correspond to the actual angular velocities and the dashed lines correspond to their estimates.

where, $f(x,u) = \begin{bmatrix} \dot{q} \\ M^{-1}(q)(\tau - C(q,\dot{q}) + G(q) + \tau_F) \end{bmatrix}$, $T_F = \begin{bmatrix} 0 \\ M^{-1}\tau_F \end{bmatrix}$, and n is the number of joints of the manipulator.

4.4.2 Experimental Setup

In this section, the test-bed used in this research is described in some details.

4.4.2.1 Real-Time Implementation

The original robot controller for the PUMA 560 was replaced by an open architecture controller. A TRC-041 cable card set from Mark-V Automation Corp. [120] was used to provide tap points to all low-level signals inside the controller. A Multi-Q multipurpose data acquisition card from Quanser Consulting was chosen to inter-

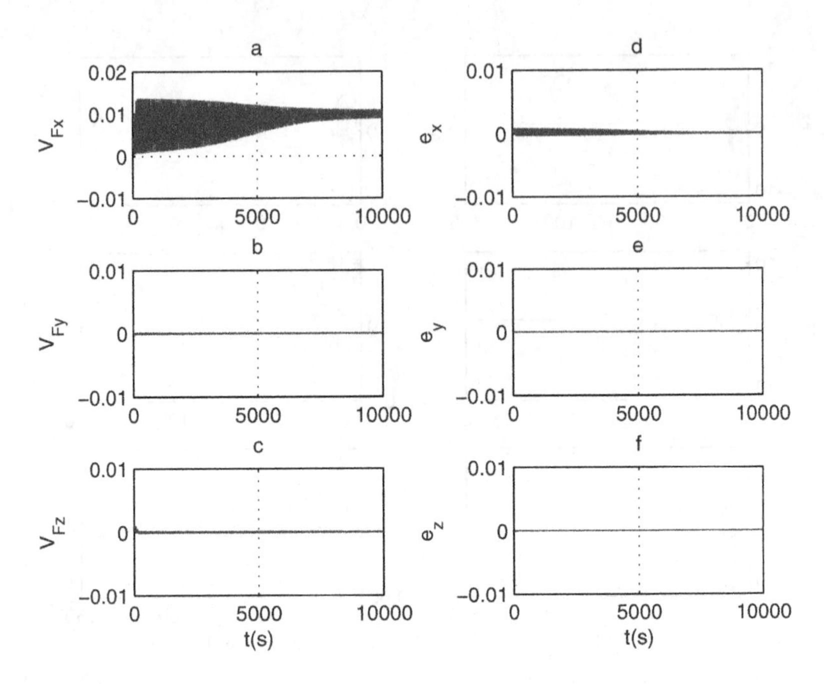

Fig. 4.6 Angular velocities of the satellite to $0.1 + 0.1sin(0.1t)$ reference trajectory with 0.01 constant fault in τ_x: (a)-(c) estimated faults; (d)-(f) angular velocity estimation errors , e_x, e_y, e_z. The solid lines correspond to the actual angular velocities and the dashed lines correspond to their estimates.

face the robot controller with the control computer, an Intel Pentium III 500Mhz PC running the VxWorks RTOS [121].

In order to communicate with the data acquisition cards, low level software drivers were written. The Simulink block diagram of the fault detection algorithm was then compiled using the Real-Time Workshop [122] to generate the C code, which runs in real-time on the control computer. A sampling rate of 100Hz was chosen for the experiments. In order to incorporate safety features in the system, a microcontroller card was designed to monitor the emergency switch and control the brakes in case of an emergency.

The numerical values of kinematic and dynamic parameters used were given in [123]. During the experiment, a significant mismatch between the responses of the mathematical model of the manipulator given by (4.17) and those of the actual manipulator was observed. The mismatch was traced to a significant amount of friction in some joints of the manipulator as well as other unmodeled dynamics which are not included in (4.17). To resolve this issue, a neural network was employed to identify the unmodeled dynamics of the robot in its normal operating condition. In fact, the *fault-free* dynamic equations of the manipulator can now be expressed as

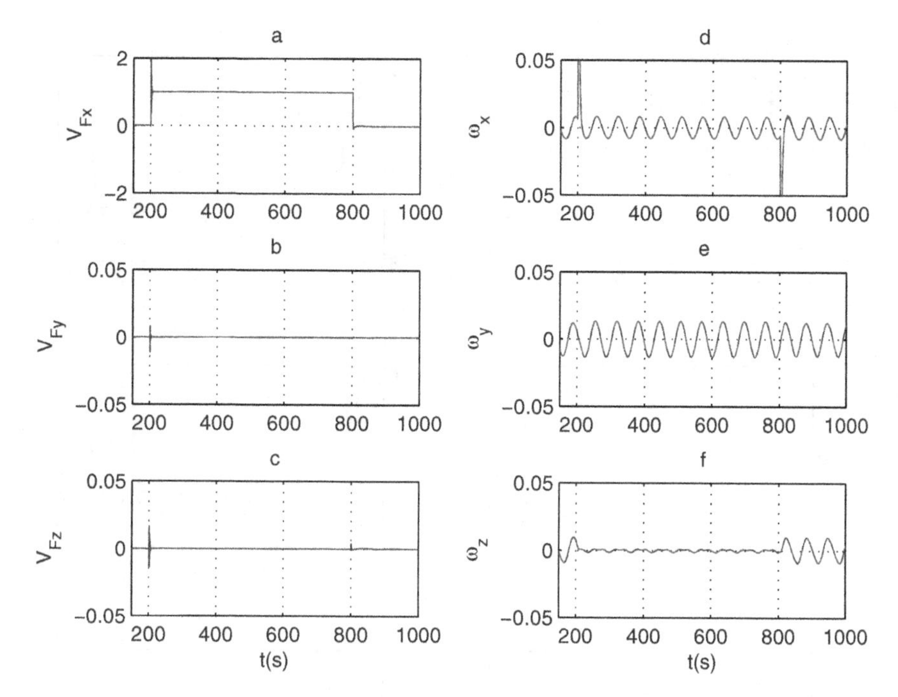

Fig. 4.7 Angular velocities of the satellite to $0.1+0.1sin(0.1t)$ reference trajectory with $1.0N.m$ constant fault in τ_x from $t = 200$ *second* to $t = 800$ *second*: (a)-(c) estimated faults; (d)-(f) satellite angular velocities, ω_x, ω_y, ω_z. The solid lines correspond to the actual angular velocities and the dashed lines correspond to their estimates.

$$M(q)\ddot{q}+C(q,\dot{q})+G(q)+\tau_N=\tau,$$

where τ_N is the signal provided by the neural network to account for unmodeled dynamics

$$\tau_N = N_1(\theta,\dot{\theta},\tau).$$

First, the robot was moved from its home position by a reference trajectory $\theta_{ref} = \theta_H + 0.3sin(t)$, where $\theta_H = [0\ -\frac{\pi}{2}\ -\frac{\pi}{2}\ 0\ 0\ 0]$ is the robot home position. A PD controller and a gravity term were used to control the robot

$$\tau_C = Kp(\theta_{ref}-\theta)+Kv(\dot{\theta}_{ref}-\dot{\theta})+G. \tag{4.19}$$

The plots of the joint positions and velocities corresponding to the actual robot and the model are shown in Fig. 4.10. As can be seen, a good match is obtained by adding τ_N to (4.17). Fig. 4.11 shows τ_N for all 6 joints of the manipulator. Next, a constant actuator fault of 5 $N-m$ was incorporated in the first three joints of the robot, i.e., $\theta_F = [5\ 5\ 5\ 0\ 0\ 0]$ and the algorithm developed in Section 4.2 was em-

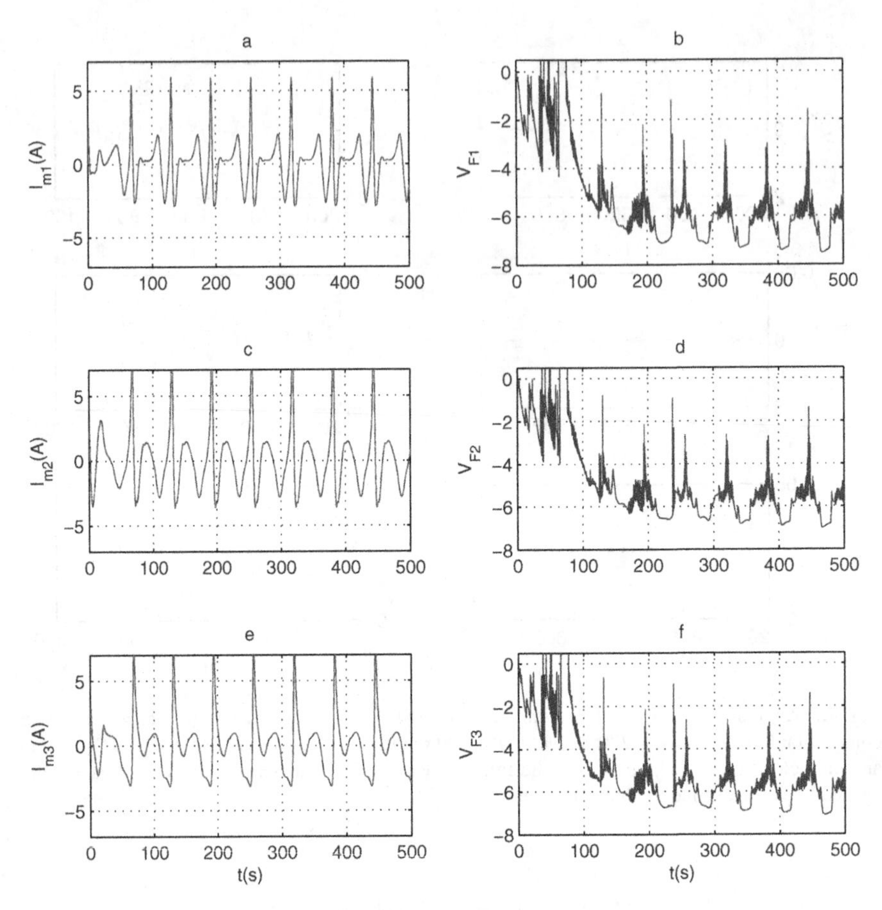

Fig. 4.8 Motor current of the reaction wheel to $0.5 sin\ 0.1t$ reference trajectory with 15.0 V constant fault in the bus voltage without control reconfiguration: (a)-(c) motor currents in x, y and z axes; (d)-(f) estimated faults. The solid lines correspond to the actual actual currents and the dashed lines correspond to their estimates.

ployed to identify the fault. The neural network has three layers, the input layer has 12 neurons, the hidden layer has 10 neurons with tangent hyperbolic activation functions, and the output layer has 6 neurons with linear transfer functions. The inputs of the network are θ and $\dot{\theta}$ and the output is the estimated fault $\hat{\tau}_F$. The responses of the fault detection system are shown in Fig. 4.12. It can be observed that the output of the neural network jumps to a nonzero torque indicating that there is a fault in the system. In fact, the neural network was able to estimate the magnitude of the fault very accurately. The plots of the joint positions and velocities are also shown in Fig. 4.13. The figures show that the states of the model follow the corresponding states of the system once the fault has been estimated by the neural network.

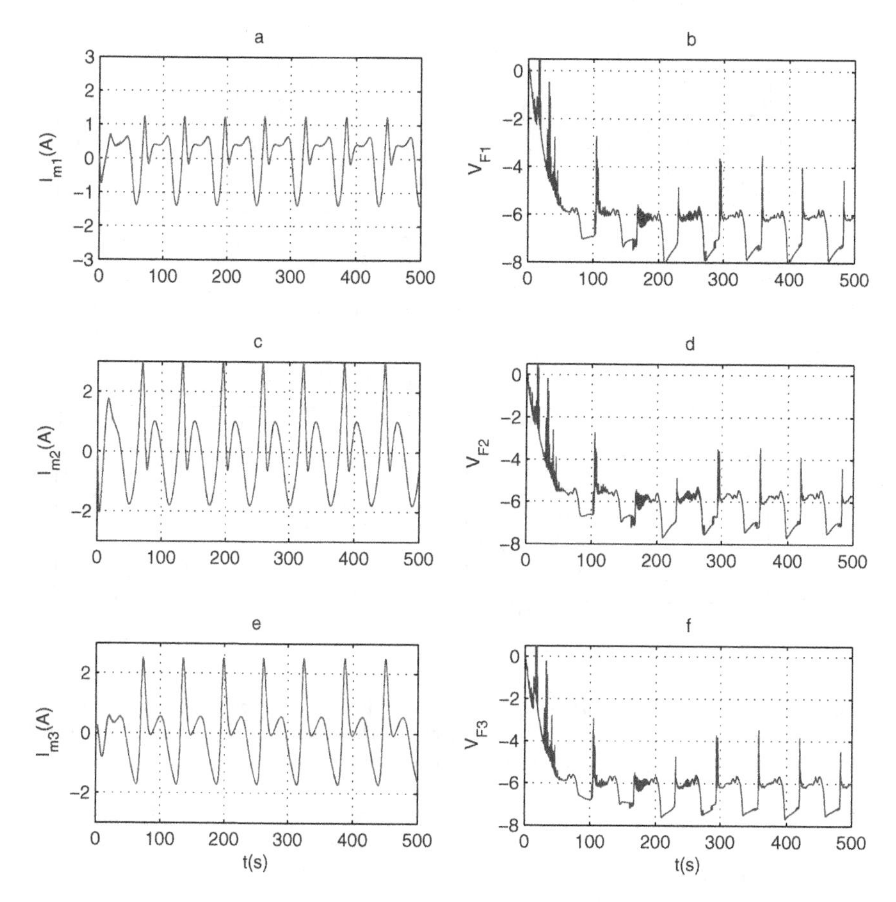

Fig. 4.9 Motor current of the reaction wheel to $0.5 sin\, 0.1t$ reference trajectory with 15.0 V constant fault in the bus voltage with control reconfiguration: (a)-(c) motor currents in x, y and z axes; (d)-(f) estimated faults. The solid lines correspond to the actual actual currents and the dashed lines correspond to their estimates.

In the next step, the output torque of the neural network (the estimated fault) was used to reconfigure the controller. In fact, the controller is now given by $\tau_C + \hat{\tau}_F$. The joint tracking errors are shown in Fig. 4.14-a to 4.14-c for the case with control reconfiguration and Fig. 4.14-d to 4.14-f for the case without control reconfiguration. It can be observed that considerable improvement is obtained when the control reconfiguration is used. These experiments confirm that the neural network was able to recognize the existence as well as the type of the fault inserted in the system.

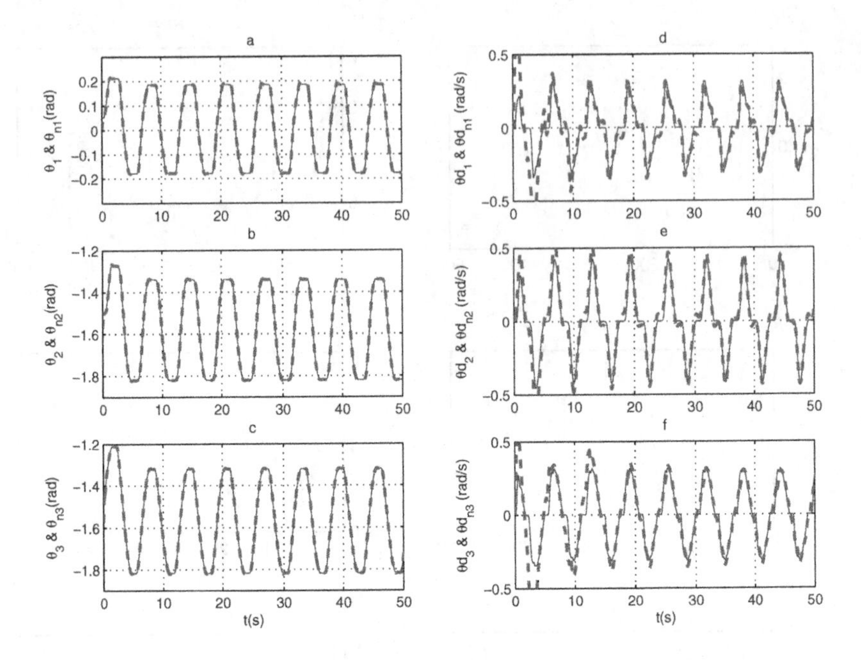

Fig. 4.10 Joint positions and velocities resulting from $0.3sin(t)$ reference trajectory in fault-free operation: (a)-(c) joint positions; (d)-(f) joint velocities. The solid lines correspond to the actual measurements and the dashed lines correspond to their estimates.

4.5 Conclusions

A neural network-based fault detection scheme was presented in this chapter. A state space approach is used and a nonlinear-in-parameters neural network (NLPNN) is employed to identify the unknown fault. The FDI scheme is based on a hybrid model of the nonlinear system composed of an analytical nominal model and a neural network model.The proposed approach has been applied via simulation to a satellite attitude control system and was also implemented on a PUMA 560 experimental test-bed.

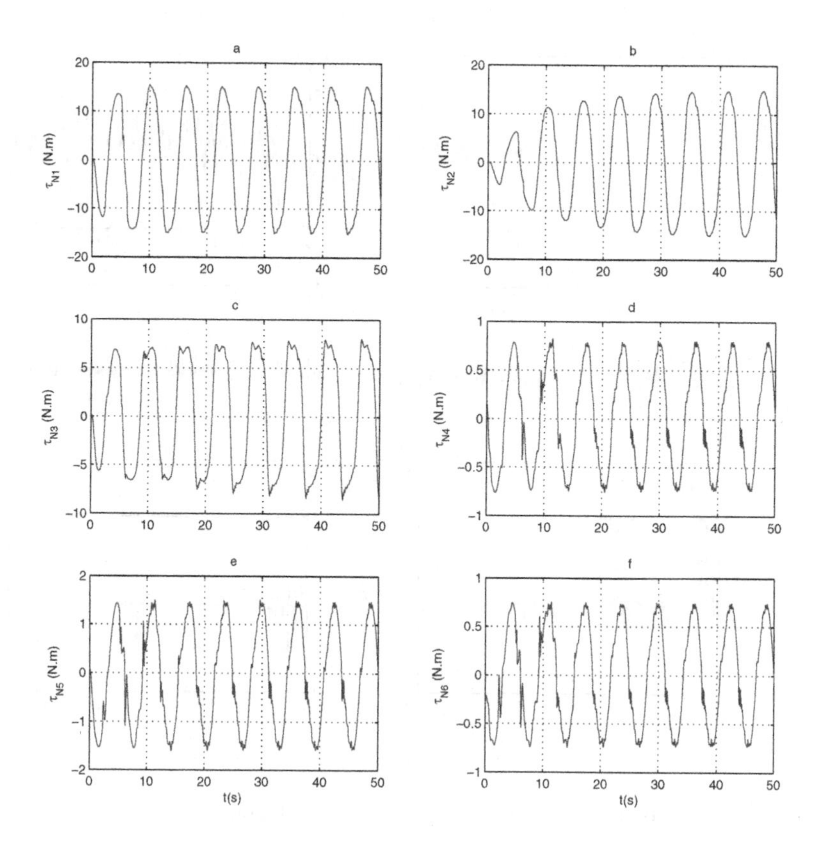

Fig. 4.11 The outputs of the neural network estimating unmodeled dynamics. (a)-(f) $\tau_{N1} - \tau_{N6}$.

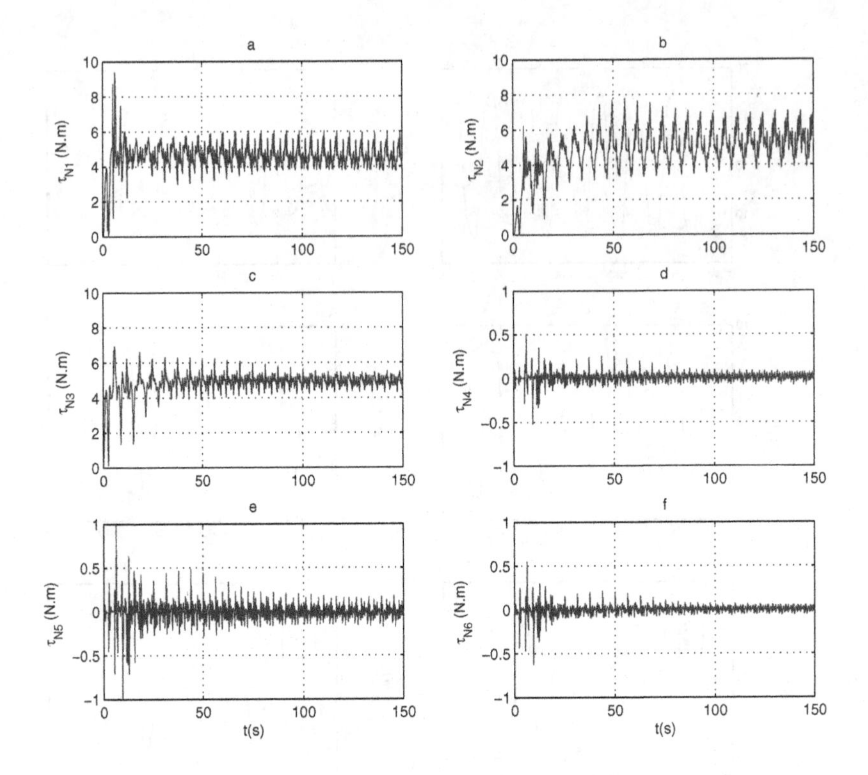

Fig. 4.12 The outputs of the neural network estimating the faults, (a)-(f) $\tau_{F1} - \tau_{F6}$.

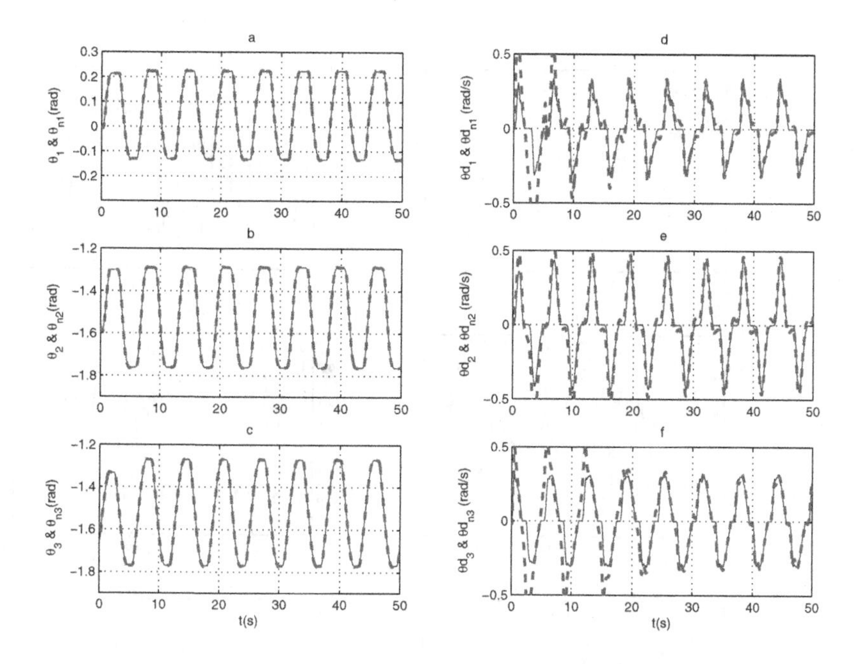

Fig. 4.13 Joint positions and velocities resulting from $0.3sin(t)$ reference trajectory with $5\ N-m$ fault in the first three actuators: (a)-(c) joint positions; (d)-(f)joint velocities. The solid lines correspond to the actual measurements and the dashed lines correspond to their estimates.

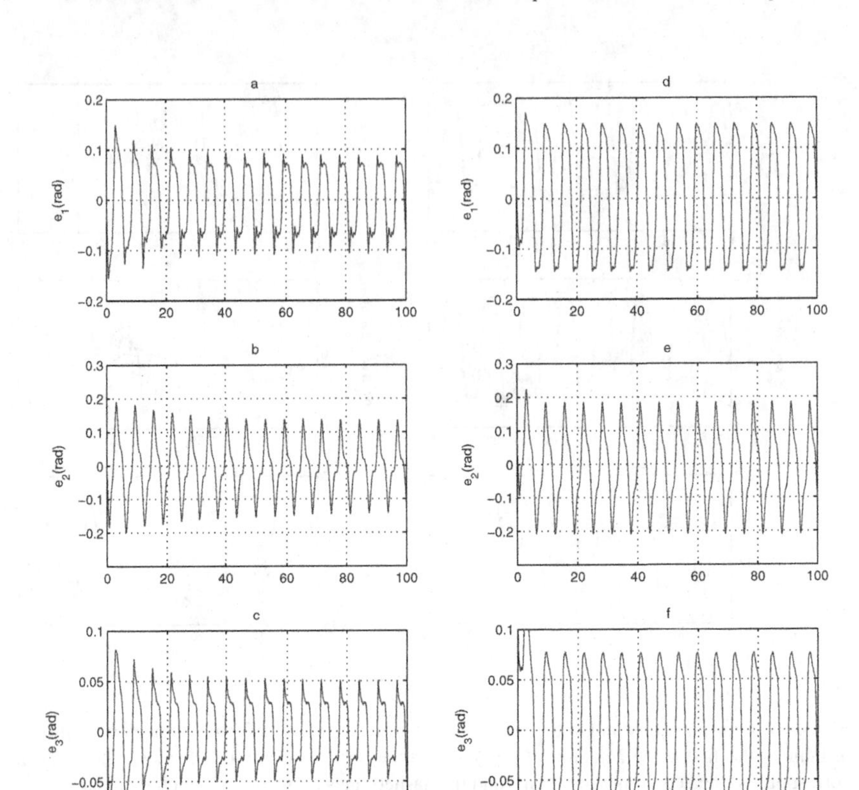

Fig. 4.14 Joint tracking errors resulting from $0.3sin(t)$ reference trajectory with $5\ N-m$ fault in the first three actuators: (a)-(c) with control reconfiguration; (d)-(f) without control reconfiguration.

Chapter 5
A Robust Actuator Gain Fault Detection and Isolation Scheme

5.1 Introduction

In this chapter, a robust fault detection and isolation strategy is presented for actuator gain fault in nonlinear systems. Unlike the FDI scheme introduced in Chapter 4 and many other methods currently exist in the literature, the proposed FDI scheme in this chapter does not rely on the availability of all state measurements. Moreover, the learning rule employed for neural network observer is based on a modified *dynamic* backpropagation methodology as opposed to static backpropagation which makes the stability analysis of the overall system much more challenging.

Several approaches have been proposed for automated fault diagnosis of nonlinear systems as summarized in [16]. Sensor and actuator fault detection and identification for nonlinear systems were considered in [16, 17, 18, 19, 20, 124]. In [125, 124], a neural network-based fault detection scheme was developed for reaction wheel type actuator. However, the algorithm assumes the availability of all states and the exact knowledge of the plant healthy behavior. Neural network-based fault detection schemes for satellite attitude control subsystems were also presented in [126, 127]. For the neural network-based fault detection schemes introduced in [114, 115, 116, 117] for satellite attitude control also it was assumed that all states are available for measurements. Moreover, stability analysis relied on the convergence of a EKF which imposes certain restrictive assumptions to hold.

It should be noted that in all of the above methods the fault functions act as an additive (bias) term on actuator signal which make them suitable for actuator *bias* fault. However, an equally important type of actuator fault is actuator **gain** fault in which the fault function acts as a multiplicative term on control (actuator) signal. Although, the intelligent algorithm developed for bias fault can also detect the gain fault, the fault severity cannot be correctly estimated. Hence, a proper fault recovery cannot be initiated based on the estimated fault. It is worth mentioning that multiplicative type of actuator faults makes the stability and convergence of FDI algorithm rather difficult.

H.A. Talebi et al., *Neural Network-Based State Estimation of Nonlinear Systems*, Lecture Notes in Control and Information Sciences 395,
DOI 10.1007/978-1-4419-1438-5_5,

In this chapter, a robust fault detection and isolation scheme is proposed for nonlinear systems subjected to ***actuator gain faults***. A neural network-based observer is used to identify the general unknown faults. The neural network weights are updated based on a *modified dynamic backpropagation* scheme. The proposed FDI scheme is easy to implement and relies on output measurements only. The stability of the overall fault detection approach in the presence of unknown actuator gain fault as well as plant and sensor uncertainties is shown using Lyapunov's direct method with no restrictive assumptions on the system and/or the FDI algorithm. The performance of the proposed FDI approach is evaluated via simulations performed on reaction wheel type actuators commonly utilized in satellite attitude control subsystems. Simulation results are presented to demonstrates the effectiveness of the proposed fault diagnosis strategy.

The reminder of the chapter is organized as follows: Section 5.2 formulates the problem. In Section 5.3, the neural network-based fault detection and isolation scheme is introduced and a detailed stability analysis is provided. The case study dealing with fault detection and isolation for the satellite's attitude control subsystem with reaction wheel type actuators introduced in Section 4.3 is applied for simulation in Section 5.5 and Section 5.6 concludes the chapter.

5.2 Problem Statement

Consider the nonlinear system

$$\begin{aligned} \dot{x}(t) &= f(x(t),u(t)) + \eta_x(x,u,t) \\ y(t) &= Cx(t) + \eta_y(x,u,t), \end{aligned} \tag{5.1}$$

where $u \in \Re^m$ is the input, $y \in \Re^m$ is the output, $x \in \Re^n$ is the state vector, $f : \Re^n \times \Re^m \longrightarrow \Re^n$ is the vector-valued known nonlinear function, $\eta_x : \Re^n \times \Re^m \times \Re \longrightarrow \Re^n$ represents the plant unmodeled dynamics and disturbances. Let $T_A(x(t),u(t),t) = diag\{T_{Ai}(x(t),u(t),t)\}$, $i = 1,...,m$ be the matrix of unknown actuator gain faults. Hence, $u(t)$ is replaced by $u_f(t) = T_A(x(t),u(t),t)u(t)$ and the nonlinear system (5.1) can be expressed as

$$\begin{aligned} \dot{x}(t) &= f(x(t),u_f(t)) + \eta_x(x,u,t) \\ y(t) &= Cx(t) + \eta_y(x,u,t). \end{aligned} \tag{5.2}$$

By adding Ax to and subtracting it from (5.2), one can get

$$\begin{aligned} \dot{x} &= Ax + g(x,u_f) + \eta_x(x,u,t) \\ y &= Cx + \eta_y(x,u,t), \end{aligned} \tag{5.3}$$

where A is a Hurwitz matrix and $g(x,u_f) = f(x,u_f) - Ax$. Note that in healthy operation, $T_A(x(t),u(t),t) = 1$, hence $u_f(t) = u(t)$. For the sake of simplicity the time dependency of the system variables is not explicitly shown in the above equation.

The following assumptions are made in order to facilitate the design and analysis of our proposed fault detection scheme.

Assumption 5.1 *The nonlinear system (5.3) is observable.*

Assumption 5.2 *The nominal closed-loop system is stable, i.e., there exists an output feedback control function* $u = k(y) = k(Cx)$ *such that* $\dot{x} = f_k(x) = f(x,k(Cx))$ *is asymptotically stable.*

Assumption 5.3 *The plant and sensor uncertainty vectors* η_x *and* η_y *are uniformly bounded, i.e.,*

$$\begin{aligned} \|\eta_x\| &\leq \bar{\eta}_x \\ \|\eta_y\| &\leq \bar{\eta}_y. \end{aligned} \tag{5.4}$$

Assumption 5.4 *Let* $\bar{x} = \begin{bmatrix} x \\ u \end{bmatrix}$. *The function* $g(x,u) = g(\bar{x})$ *is Lipschitz in* x *and* u *with Lipschitz constant* l_g, *i.e.,*

$$\|g(x,u) - g(\hat{x},\hat{u})\| \leq l_g \left\| \begin{bmatrix} x - \hat{x} \\ u - \hat{u} \end{bmatrix} \right\|. \tag{5.5}$$

Now, the fault detection and isolation problem can be formally stated as:
Objective: Given Assumptions 5.1 to 5.4, develop a methodology that can detect, isolate, and estimate the fault vector T_A in a stable fashion.

Note that if the fault vector is estimated correctly, the isolation and detection objective are automatically satisfied. In the following section, a neural network-based methodology is presented for this purpose. One of the main advantages of the proposed scheme is that the fault detection, isolation, and estimation steps are all unified. That is to say, the signal generated by the neural networks serve as residuals as well as estimated faults in corresponding actuator. Hence, neither extra measured/calculated signals nor a separate fault isolation policy is required to isolate the fault.

5.3 Neural Network-Based Fault Detection and Estimation Scheme

According to the representation (5.3), recurrent network models are now constructed by parameterizing the maps T_A through utilization of feedforward neural network architecture, denoted by N_1. Specifically, the following model is considered for our observer-based design strategy

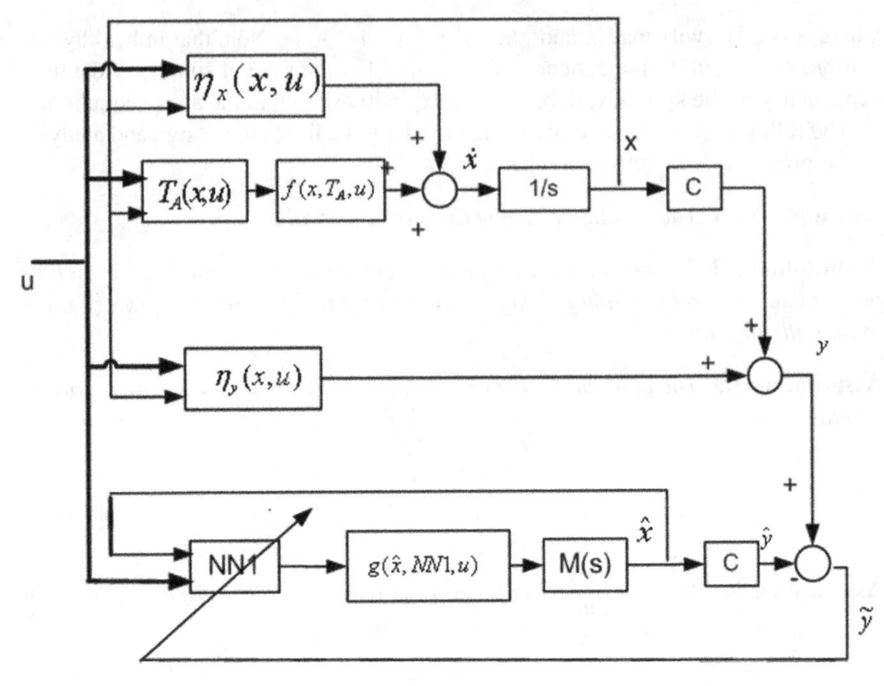

Fig. 5.1 Structure of the neural network-based fault detection and isolation scheme.

$$\begin{aligned}\dot{\hat{x}} &= A\hat{x} + g(\hat{x}, \hat{u}_f)\\ \hat{u}_f &= \hat{T}_A(\hat{x}, u, t)u\\ \hat{y} &= C\hat{x},\end{aligned} \tag{5.6}$$

where $\hat{T}_A(\hat{x}, u, W)$ is the map performed by the neural network N_1. The structure of the estimator is shown in Fig. 5.1.

Based on Theorem 1.1, a recurrent nonlinear-in-parameters neural network (NLPNN) can represent nonlinear function $T_A(x,u,t)$ as follows:

$$T_A(x,u,t) = W\sigma(V\bar{x}) + \varepsilon(\bar{x}),$$

where $\bar{x} = \begin{bmatrix} x \\ u \end{bmatrix}$, W, and V are the neural networks ideal weight matrices, $\varepsilon(\bar{x}) \leq \varepsilon_N$ is the neural network bounded approximation error, and $\sigma(.)$ is the hidden layer transfer function which in this chapter it is considered as a tangent hyperbolic function introduced in (2.4).

Therefore, the function $T_A(x,u,W,t)$ can be approximated as

$$\hat{T}_A(\hat{\bar{x}}, W, t) = \hat{W}\sigma(\hat{V}\hat{\bar{x}}), \tag{5.7}$$

where $\hat{\bar{x}} = \begin{bmatrix} \hat{x} \\ u \end{bmatrix}$. The estimator model can now be written as

$$\begin{aligned} \dot{\hat{x}}(t) &= A\hat{x} + g(\hat{x}, \hat{u}_f) \\ \hat{u}_f &= \hat{W}\sigma(\hat{V}\hat{\bar{x}})u \\ \hat{y} &= C\hat{x}. \end{aligned} \tag{5.8}$$

Denoting the state estimation error as $\tilde{x} = x - \hat{x}$, the output estimation error as $\tilde{y} = y - \hat{y}$, and by using (5.3) and (5.8), the error dynamics can be stated as follows:

$$\begin{aligned} \dot{\tilde{x}}(t) &= A\tilde{x} + g(x, u_f) - g(\hat{x}, \hat{u}_f) + \eta_x(\bar{x}) \\ \tilde{y} &= C\tilde{x} + \eta_y(\bar{x}), \end{aligned} \tag{5.9}$$

where $\tilde{x} = x - \hat{x}$.

5.3.1 Fault Detection and Isolation Policy

The neural networks employed in our work are to identify the fault and produce an identity matrix output in non-faulty (healthy) operation. When there is a fault in the actuator, the output of the neural network NN_1 i.e., $\hat{T}_A$ deviates from unity and that is used as a residual to detect and estimate the severity of the fault. An actuator fault is declared if the residual ($\hat{T}_A$) exceeds a certain threshold value.

5.3.2 Stability Analysis

In this section, the learning rules for updating the weights of the neural networks are given and the stability of the fault detection and isolation scheme is shown by utilizing Lyapunov's direct method. The main result is stated in the following theorem.

Theorem 5.1. *Consider the nonlinear system (5.2) and the observer model (5.6). Given Assumptions 5.1 to 5.4, if the weights of the neural network are updated according to:*

$$\dot{\hat{W}} = -\eta_1\left(\frac{\partial J}{\partial \hat{W}}\right) - \rho_1 \|\tilde{y}\| \hat{W} \tag{5.10}$$

$$\dot{\hat{V}} = -\eta_2\left(\frac{\partial J}{\partial \hat{V}}\right) - \rho_2 \|\tilde{y}\| \hat{V}, \tag{5.11}$$

then $\tilde{x}, \tilde{y}, \tilde{W}$, and $\tilde{V} \in L_\infty$, i.e., the state and output estimation errors, and the weights error are all ultimately bounded in the presence of unknown actuator faults and plant and sensor uncertainties.

Remark 5.1 *In the above equations, η_1 and η_2 are positive constants representing the learning rates, ρ_1 and ρ_2 are some small positive constants, and $J = \frac{1}{2}(\tilde{y}^T \tilde{y})$ is the neural network objective function.*

Proof: Let us define

$$net_{\hat{v}} = \hat{V}\hat{\bar{x}} \tag{5.12}$$
$$net_{\hat{w}} = \hat{W}S_1, \tag{5.13}$$

where

$$S_1 = \sigma(\hat{V}_1\hat{\bar{x}}). \tag{5.14}$$

Note that S_1 is bounded due to the boundedness of the tangent hyperbolic function $\sigma(.)$.

Next, $\frac{\partial J}{\partial \hat{W}}$ and $\frac{\partial J}{\partial \hat{V}}$ can be computed as

$$\frac{\partial J}{\partial \hat{W}} = \frac{\partial J}{\partial net_{\hat{w}}} \frac{\partial net_{\hat{w}}}{\partial \hat{W}}$$
$$\frac{\partial J}{\partial \hat{V}} = \frac{\partial J}{\partial net_{\hat{v}}} \frac{\partial net_{\hat{v}}}{\partial \hat{V}}.$$

We have

$$\frac{\partial J}{\partial net_{\hat{w}}} = \frac{\partial J}{\partial \tilde{y}} \frac{\partial \tilde{y}}{\partial \hat{x}} \frac{\partial \hat{x}}{\partial net_{\hat{w}}} = -\tilde{y}^T C \frac{\partial \hat{x}}{\partial net_{\hat{w}}}$$
$$\frac{\partial J}{\partial net_{\hat{v}}} = \frac{\partial J}{\partial \tilde{y}} \frac{\partial \tilde{y}}{\partial \hat{x}} \frac{\partial \hat{x}}{\partial net_{\hat{v}}} = -\tilde{y}^T C \frac{\partial \hat{x}}{\partial net_{\hat{v}}}, \tag{5.15}$$

and

$$\frac{\partial net_{\hat{w}}}{\partial \hat{W}} = S_1^T \tag{5.16}$$
$$\frac{\partial net_{\hat{v}}}{\partial \hat{V}} = \hat{\bar{x}}^T. \tag{5.17}$$

Now, by using (5.8) and the definitions of $net_{\hat{v}}$ and $net_{\hat{w}}$ as (5.12) and (5.13) respectively, we can write

$$\frac{\partial \dot{\hat{x}}(t)}{\partial net_{\hat{w}}} = A\frac{\partial \hat{x}}{\partial net_{\hat{w}}} + \frac{\partial g}{\partial \hat{x}} \frac{\partial \hat{x}}{\partial net_{\hat{w}}} + \frac{\partial g}{\partial \hat{u}_f} \frac{\partial \hat{u}_f}{\partial net_{\hat{w}}} \tag{5.18}$$
$$\frac{\partial \dot{\hat{x}}(t)}{\partial net_{\hat{v}}} = A\frac{\partial \hat{x}}{\partial net_{\hat{v}}} + \frac{\partial g}{\partial \hat{x}} \frac{\partial \hat{x}}{\partial net_{\hat{v}}} + \frac{\partial g}{\partial \hat{u}_f} \frac{\partial \hat{u}_f}{\partial net_{\hat{v}}}. \tag{5.19}$$

Now, by using (5.8), we have

$$\frac{\partial \hat{u}_f}{\partial net_{\hat{w}}} = u$$
$$\frac{\partial \hat{u}_f}{\partial net_{\hat{v}}} = \hat{W}(I - \Lambda(\hat{V}\hat{\bar{x}}))u,$$

where

$$\Lambda(\hat{V}\hat{\bar{x}}) = diag\{\sigma_i^2(\hat{V}_i\hat{\bar{x}})\}, i = 1,2,...,m. \tag{5.20}$$

Now, denote:

$$d_{xw} := \frac{\partial \hat{x}}{\partial net_{\hat{w}}} \tag{5.21}$$
$$d_{xv} := \frac{\partial \hat{x}}{\partial net_{\hat{v}}} \tag{5.22}$$
$$S_2 = I - \Lambda(\hat{V}\hat{\bar{x}}).$$

Note that S_2 is in fact the derivative of the tangent hyperbolic function which is also known to be bounded. Then, by changing the order of partial differentiation, we get:

$$\dot{d}_{xw} = Ad_{xw} + \frac{\partial g}{\partial \hat{x}} d_{xw} + \frac{\partial g}{\partial \hat{u}_f} u \tag{5.23}$$
$$\dot{d}_{xv} = Ad_{xv} + \frac{\partial g}{\partial \hat{x}} d_{xv} + \frac{\partial g}{\partial \hat{u}_f} \hat{W} S_2 u. \tag{5.24}$$

The above equations represent a set of nonlinear dynamical systems. Let

$$A_c = \left.\frac{\partial g}{\partial \hat{x}}\right|_{\hat{x}=0, \hat{u}_f=u}$$
$$B = \left.\frac{\partial g}{\partial \hat{u}_f}\right|_{\hat{x}=0, \hat{u}_f=u}. \tag{5.25}$$

Then, consider the linearization of the above equation around $(0, u)$:

$$\dot{d}_{xw} \approx Ad_{xw} + A_c d_{xw} + Bu \tag{5.26}$$
$$\dot{d}_{xv} \approx Ad_{xv} + A_c d_{xv} + B\hat{W} S_2 u. \tag{5.27}$$

Now, by using (5.15), (5.16), (5.17), (5.21), and (5.22) the learning rules (5.10) and (5.11) can be written as

$$\dot{\hat{W}} = \eta_1(\tilde{y}^T C d_{xw})^T S_1^T - \rho_1 \|\tilde{y}\| \hat{W} \tag{5.28}$$
$$\dot{\hat{V}} = \eta_2(\tilde{y}^T C d_{xv})^T \hat{\bar{x}}^T - \rho_2 \|\tilde{y}\| \hat{V}. \tag{5.29}$$

The learning rules (5.28) and (5.29) in terms of the weight errors $\tilde{W} = W - \hat{W}$ and $\tilde{V} = V - \hat{V}$, can be written as

$$\dot{\hat{W}} = -d_{xw}^T l_0 \tilde{y} S_1^T + \rho_1 \|\tilde{y}\| \hat{W} \tag{5.30}$$

$$\dot{\hat{V}} = -d_{xv}^T l_3 \tilde{y} \hat{\bar{x}}^T + \rho_2 \|\tilde{y}\| \hat{V} \tag{5.31}$$

$$\dot{d}_{xw} = A d_{xw} + A_c d_{xw} + Bu \tag{5.32}$$

$$\dot{d}_{xv} = A d_{xv} + A_c d_{xv} + B\hat{W} S_2 u, \tag{5.33}$$

where $l_0 = \eta_1 C^T, l_3 = \eta_2 C^T$. The learning procedure is as follows. First (5.32) and (5.33) are solved for gradients d_{xw} and d_{xv}. Then, the learning rules (5.30) and (5.31) are solved to update the weights of the network.

Now, the following observations can be made:

Fact 5.1.1 *Equation (5.25) shows that (5.26) is obtained by linearizing the nominal system (5.3) around* $(0,u)$. *Hence, given the stability of the nominal system (see Assumption 5.2), it is clear that (5.26) also represents a stable linear dynamical system with bounded input u, hence* d_{xw} *is also bounded.*

Fact 5.1.2 *The boundedness of* $\tilde{V}$ *has no effect on the stability of* $\tilde{x}$ *(5.9) and* $\tilde{W}$ *(5.30) since* $S_1 \in L_\infty$ *regardless of the value of* $\tilde{V}$.

Fact 5.1.3 *The function* $\Lambda(.)$ *defined in (5.20) is bounded due to the boundedness of the tangent hyperbolic function, hence* $S_2 \in L_\infty$.

Fact 5.1.4 *The boundedness of gradient* d_{xv} *has no effect on the stability of* $\tilde{x}$ *and* $\tilde{W}$ *since it only affects* $\tilde{V}$ *which according to Fact 5.1.2 has no effect on boundedness of (5.9) and (5.30). Moreover, the gradient* d_{xv} *(5.33) remains bounded if* $\hat{W}$ *is bounded since (5.33) represent a stable linear dynamical system with bounded input (see Fact 5.1.1).*

Given Facts 5.1.2 and 5.1.4, we can conclude that the system can be decomposed into two subsystems for the purpose of stability analysis. One subsystem (Subsystem 1) consists of the estimation error dynamics (5.9), the output-layer weight error (5.30) and the gradient d_{xw} (5.32). The other subsystem (Subsystem 2) consists of the hidden layer weight error (5.31) and the gradient d_{xv} (5.33). Note that, the two subsystems are not completely decoupled due to the presence of $\tilde{x}$ and $\hat{W}$ in (5.31) and (5.33), respectively. Consequently, the stability of Subsystem 1 can be shown independent of Subsystem 2, but not vice versa. To study the stability of Subsystem 1, we select the positive definite Lyapunov function candidate as

$$L = \frac{1}{2}\{\tilde{x}^T P \tilde{x} + tr(\tilde{W}^T \rho_1^{-1} \tilde{W})\},$$

where $P = P^T$ is a positive-definite matrix satisfying the Lyapunov equation:

$$A^T P + PA = -Q, \tag{5.34}$$

for the Hurwitz matrix A and some positive-definite matrix Q. The time derivative of L is given by

$$\dot{L} = \frac{1}{2}\dot{\tilde{x}}^T P \tilde{x} + \frac{1}{2}\tilde{x}^T P \dot{\tilde{x}} + tr(\tilde{W}^T \rho_1^{-1} \dot{\tilde{W}}). \tag{5.35}$$

Now, by substituting (5.9), (5.30), and (5.34) into (5.35), one can get

$$\begin{aligned}\dot{L} = &-\frac{1}{2}\tilde{x}^T Q\tilde{x} + \tilde{x}^T P(g(x,u_f) - g(\hat{x},\hat{u}_f) + \eta_x(\bar{x})) \\ &+ tr(-\tilde{W}^T(d_{xw}^T l_1\tilde{x} + d_{xw}^T l_2\eta_y)S_1^T + \tilde{W}^T\|C\tilde{x} + \eta_y\|(W - \tilde{W})),\end{aligned}$$

where $l_1 = \rho_1^{-1} l_0 C$ and $l_2 = \rho_1^{-1} l_0$.

Given assumption 4, we have

$$\|g(x,u_f) - g(\hat{x},\hat{u}_f)\| \le l_g \left\| \begin{bmatrix} x - \hat{x} \\ u - \hat{u} \end{bmatrix} \right\| \le l_g \left(\|\tilde{x}\| + \|u_f - \hat{u}_f\|\right). \tag{5.36}$$

Recall that $u_f = (W\sigma(V\bar{x}) + \varepsilon(\bar{x}))u$ and $\hat{u}_f = \hat{W}\sigma(\hat{V}\hat{\bar{x}})u$, hence

$$u_f - \hat{u}_f = (\tilde{W}S_1 + w(t))u, \tag{5.37}$$

where $w(t) = W[\sigma(V\bar{x}) - \sigma(\hat{V}\hat{\bar{x}})] + \varepsilon(\bar{x})$ is a bounded disturbance term [80], i.e., $\|w(t)\| \le \bar{w}$ for some positive constant $\bar{w}$, due to the characteristics of the tangent hyperbolic function.

Note that the following inequalities hold:

$$\begin{aligned} tr(\tilde{W}^T(W - \tilde{W})) &\le W_M\|\tilde{W}\| - \|\tilde{W}\|^2 \\ tr(-\tilde{W}^T d_{xw}^T l_1\tilde{x}S_1^T) &\le \sigma_m\|\tilde{W}\|\,\|d_{xw}\|\|l_1\|\,\|\tilde{x}\|, \end{aligned} \tag{5.38}$$

where W_M, V_M and σ_m are the upper bounds for tangent hyperbolic function $\sigma(.)$ and the ideal weights V and W. Please, also note that $\|d_{xw}\|$ is bounded since d_{xw} is bounded (ref. to Fact 5.1.1).

Consequently, by using (5.4), (5.5), (5.36) (5.37), and (5.38), we get

$$\begin{aligned} \dot{L} &\le F \\ F &= -\frac{1}{2}\lambda_{min}(Q)\|\tilde{x}\|^2 + \|\tilde{x}\|\|P\|(l_g\|\tilde{W}\|\sigma_m u + l_g\bar{w}u + l_g\|\tilde{x}\| + \bar{\eta}_x) \\ &+ \sigma_m\|\tilde{W}\|(\|l_1\|\bar{d}_{xw}\|\tilde{x}\| + \bar{\eta}_y\bar{d}_{xw}\|l_2\|) + (W_M\|\tilde{W}\| - \|\tilde{W}\|^2)(\|C\|\|\tilde{x}\| + \bar{\eta}_y), \end{aligned}$$

where $\lambda_{min}(Q)$ is the minimum eigenvalue of Q. Now, by rearranging F, we get:

$$\begin{aligned} F &= -\alpha_1\|\tilde{x}\|^2 + \alpha_2\|\tilde{x}\| + \alpha_3\|\tilde{x}\|\|\tilde{W}\| + \alpha_4\|\tilde{W}\| \\ &- \alpha_5\|\tilde{W}\|^2\|\tilde{x}\| - \alpha_6\|\tilde{W}\|^2 \\ \alpha_1 &= \frac{1}{2}\lambda_{min}(Q) - l_g\|P\|,\ \alpha_2 = (l_g\bar{w}\bar{u} + \bar{\eta}_x)\|P\| \\ \alpha_3 &= \sigma_m(l_g\bar{u}\|P\| + \bar{d}_{xw}\|l_1\|) + W_M\|C\| \\ \alpha_4 &= \bar{\eta}_y(\sigma_m\bar{d}_{xw}\|l_2\| + W_M),\ \alpha_5 = \|C\|,\ \alpha_6 = \bar{\eta}_y, \end{aligned} \tag{5.39}$$

where $\bar{u}$ and $\bar{d}_{xw}$ are the upper bounds of u and d_{xw}, respectively. Note that all parameters defined above are positive except α_1 which can be kept positive if $\lambda_{min}(Q) > 2l_g\|P\|$. Also, note that u is bounded since the original system is assumed to be stable (Assumption 5.2). Now, by completing the squares for the terms involving $\|\tilde{W}\|$, or $\|\tilde{x}\|$ we try to find some conditions on $\|\tilde{x}\|$ or $\|\tilde{W}\|$, that ensure negative definiteness of the derivative of the Lyapunov function candidate. Now, by performing some algebraic manipulations, we obtain the conditions that ensure the negative definiteness of $\dot{L}$. In fact, provided that $\lambda_{min}(Q) > 2l_g\|P\|$, $\dot{L} < 0$ if

$$\|\tilde{x}\| > x^0 = \frac{(\alpha_2+\beta_3)+\sqrt{(\alpha_2+\beta_3)^2+4\beta_4\alpha_1}}{2\alpha_1}, \tag{5.40}$$

Or

$$\|\tilde{W}\| > w^0 = \frac{\alpha_4+\sqrt{\alpha_4^2+4\alpha_6\beta_6}}{2\alpha_6}, \tag{5.41}$$

where $\beta_3 = \frac{\alpha_3^2}{4\alpha_5}$, $\beta_4 = \frac{\alpha_4^2}{4\alpha_6}$, $\beta_5 = \alpha_2 + \frac{\alpha_3^2}{4\alpha_5}$, and $\beta_6 = \frac{\beta_5^2}{4\alpha_1}$.

Equations (5.40) and (5.41) specify the balls χ_x and χ_w according to

$$\chi_x = \{\tilde{x} \mid \|\tilde{x}\| > x^0\}$$
$$\chi_w = \{\tilde{W}_2 \mid \|\tilde{W}_2\| > w_2^0\}.$$

In fact, $\dot{L}$ is negative definite outside the balls χ_x and χ_w. This shows that the estimation error $\tilde{x}$ and the weight error $\tilde{W}$ cannot leave the balls χ_x and χ_w, respectively, since the growth of $\tilde{x}$ or $\tilde{W}$ outside the balls χ_x or χ_w results in the negativeness of $\dot{L}$ and hence the reduction of L which in turn results in the reduction of its argument $\tilde{x}$ and $\tilde{W}$. Hence, we can conclude that $\tilde{x}$ and $\tilde{W}$ are ultimately bounded. The ultimate boundedness of $\tilde{y}$ immediately follows from (5.9) since $\tilde{x}$ and $\eta_y(x)$ are bounded.

The above analysis proves the stability of Subsystem 1. To show the ultimate boundedness of $\tilde{V}$ and d_{xv}, consider (5.31) and (5.33) which can be expressed as

$$\begin{aligned}\dot{\tilde{V}} &= f_2(\tilde{y}, d_{xv}, \hat{\bar{x}}) + \rho_2\|\tilde{y}\|\hat{V} \\ &= f_2(\tilde{y}, d_{xv}, \hat{\bar{x}}) + \gamma_2 V - \gamma_2 \tilde{V} \qquad (5.42)\\ \dot{d}_{xv} &= Ad_{xv} + A_c d_{xv} + B\hat{W}S_2 u, \qquad (5.43)\end{aligned}$$

where

$$f_2(\tilde{y}, d_{xv}, \hat{\bar{x}}) = d_{xv}^T l_2 \tilde{y} \hat{\bar{x}}^T$$
$$\gamma_2 = \rho_2\|\tilde{y}\|.$$

Having shown the boundedness of $\tilde{W}$ and given Fact 5.1.3, it can be clearly seen from (5.43) that the gradient d_{xv} is also bounded, since it represents a stable linear system subject to a bounded input $B\hat{W}S_2u$. Now, given that $d_{xv} \in L_\infty$, it can be observed that $f_2(.)$ is also bounded since $\tilde{y}$ and S_1 are both bounded, C is bounded and d_{xw} is also bounded. Given the fact that the ideal weight V is fixed, (5.42) can be regarded as a linear system with bounded input $(f_2(\tilde{y}, d_{xv}, \hat{\bar{x}}) + \gamma_2 V)$. It is clear that this system is stable since γ_2 is positive and the system input remains bounded.

Hence, ultimate boundedness of $\tilde{V}$ is also ensured. This completes the proof of the theorem. □

5.4 A Case Study: Application to a Satellite's Attitude Control Subsystem

In this section, the fault detection and isolation scheme developed in the previous section is applied to reaction wheel type actuators used in the satellites attitude control subsystem which the dynamics were given in Section 4.3.

Now, let us recall the dynamics of reaction wheel illustrated in Fig. 4.2:

$$\begin{bmatrix} \dot{I}_m \\ \dot{\omega}_m \end{bmatrix} = \begin{bmatrix} G_d\omega_d\left[\Psi_1(I_m,\omega_m) - \Psi_3(\omega_m)\right] - \omega_d I_m \\ \frac{1}{J}\left[K_t I_m - \tau_c \Psi_2(\omega_m) - \tau_v \omega_m\right] \end{bmatrix} + \begin{bmatrix} G_d\omega_d \\ 0 \end{bmatrix} V_{com}$$
$$\tau = K_t I_m, \tag{5.44}$$

where Ψ_1, Ψ_2, and Ψ_3 represent the nonlinearities for EMF torque limiting, Coulomb friction, and speed limiter subsystems, respectively. The structure of the **unknown fault** in this chapter is considered as a multiplicative term on the command voltage V_{com} in reaction wheel dynamic equation (5.44):

$$\begin{aligned} \dot{I}_m &= G_d\omega_d\left[\Psi_1(I_m,\omega_m) - \Psi_3(\omega_m)\right] - \omega_d I_m \\ &\quad + G_d\omega_d(T_A(V_{com},\Omega,\omega,I_m,\omega_m)V_{com}), \end{aligned} \tag{5.45}$$

where T_A represents the unknown fault. Now, by defining $x = \begin{bmatrix} \Omega \\ \omega \\ I_m \\ \omega_m \end{bmatrix}$ and $u = V_{com}$,

Equations (4.10), (4.12), (5.45), and (5.44) can be written as

$$\dot{x} = f(x, T_A(\bar{x})u). \tag{5.46}$$

5.5 Simulation Results

For the purpose of simulation, the inertia matrix and the reaction wheel parameters given in (4.16) and Table 4.1 are used in this example. A conventional state feedback controller is used to stabilize the closed-loop system in fault-free operation. All simulation results are obtained by incorporating 20% and 30% **uncertainties** in Coulomb friction constant τ_c and motor torque constant K_t, respectively. All sensory measurements are assumed to be corrupted with 10% Gaussian noise. The neural networks approximating actuator faults, i.e., NN_1 has three layers where the input layer has 15 neurons, the hidden layer has 5 neurons with tangent hyperbolic activation functions, and the output layer has 3 neurons with linear transfer functions. The input of both networks is $\hat{\bar{x}}$ as defined in Section 5.3 and is shown in Fig. 5.1. The output of NN_1 is $\hat{T}_A = [\hat{T}_{A1}\ \hat{T}_{A2}\ \hat{T}_{A3}\ ]^T$. The numerical values used for the neural networks learning parameters in (5.10) and (5.11) are: $\eta_1 = \eta_2 = 1; \rho_1 = \rho_2 = 10^{-6}$, and the Hurwitz matrix A is selected as $A = -2I_{12}$, where I_{12} denotes a 12×12 identity matrix.

First, a non-faulty situation was considered and a $0.15sin(0.01t)$ input trajectory was applied to the satellite. The estimated as well as actual states (namely, reaction wheel motor currents) and the estimated faults are shown in Fig. 5.2. As can be seen, despite the noisy and corrupted sensory measurements, the magnitudes of the neural network outputs approach to a value close to one confirming that there is no fault in the system. Note that the outputs of the networks do not approach to one due to the presence of measurement noises and uncertainties in reaction wheel parameters. Note also that, care should be taken in selecting an appropriate value of the threshold. Based on the upper bounds assumed for uncertainties and measurement noises (see (5.4)), the threshold value was selected as 0.45 in our simulations. The fault detection policy is stated as follows: Let t_f be the time when a residual exceed the corresponding threshold value. Then, an actuator/sensor fault is declared if

$$
\begin{aligned}
&|\hat{T}_{Ai}| > th_{ai},\ i = 1,\ \dots 3,\\
&\text{Or}\\
&|\hat{T}_{Si}| > th_{si},\ i = 1,\ \dots 3\\
&\forall t \in [t_f, t_f + t_r],
\end{aligned}
$$

where th_{ai} and th_{si} are the threshold values corresponding to the ith actuator/sensor, respectively. Furthermore, t_r is selected in the above equation, to allow the output of neural networks reach their steady state responses. In our simulations, t_r is selected as $t_r = 70\ second$.

Next, a 20% gain fault was introduced exclusively for channel y reaction wheel for the time period $t \in [100\ 200]\ second$, i.e.,

$$
T_A = \begin{cases} [1\ 0.8\ 1]^T & t \in [100\ 200] \text{ second} \\ [1\ 1\ 1]^T & \text{otherwise} \end{cases}.
$$

The responses of the fault detection and isolation system are shown in Fig. 5.3. It can be observed that **only** the second element of the neural network output, i.e., T_{A2} approaches to a non-unity value corresponding very closely to the actual magnitude of the fault that was injected to only the channel y actuator. These responses indicate that there is a fault in channel y actuator confirming the excellent isolation property of the proposed FDI approach. To exhibit the signature of the injected fault on reaction wheel angular velocities, the response of ω_m are also shown in Fig. 5.3-a to 5.3-c. A change in ω_{m2} at $t = 100\ second$ (see Fig. 5.3-b) demonstrates the effect of the actuator fault.

Next, the ability of the proposed fault detection and isolation scheme to identify simultaneous faults as well as detecting and estimating small gain faults was evaluated. Towards this end, constant faults are introduced in both channel x and channel z actuators for the time period $t \in [100\ 200]$ *second*, i.e.,

$$T_A = \begin{cases} [0.95\ 1\ 0.7]^T & t \in [100\ 200] \text{ second} \\ [1\ 1\ 1]^T & \text{otherwise} \end{cases}.$$

The results are shown in Fig. 5.4-a to 5.4-f. Once again, it can be seen that the first and third elements of the network approach the true values of the faults inserted in the system while the second output of NN_1 practically remains at unity. To exhibit the signature of the injected fault on reaction wheel angular velocities, the response of ω_{m1} is enlarged on the event of fault occupance, i.e., as shown in Fig. 5.4-a. It can be seen that despite the little effect of this small fault on system responses, the neural network could detect, isolate, and actually estimate the magnitude of the injected fault.

5.6 Conclusions

In this chapter, a robust fault detection and isolation scheme has been proposed by employing a neural network-based observer which can detect and also identify the severity of *actuator gain* faults unlike many other method which only consider the bias fault. The methodology also considers the presence of disturbances and uncertainties in model and sensory measurements. The neural network weights update is based on a modified *dynamic* backpropagation scheme. Moreover, the proposed FDI scheme does not rely on the availability of all state measurements. The stability of the overall fault detection approach in the presence of unknown actuator gain fault as well as plant and sensor uncertainties has been shown using Lyapunov's direct method with no restrictive assumptions on the system and/or the FDI algorithm. The performance of the proposed FDI approach was evaluated via simulations performed on reaction wheel type actuators commonly utilized in the Attitude control Subsystems (ACS) of satellites.

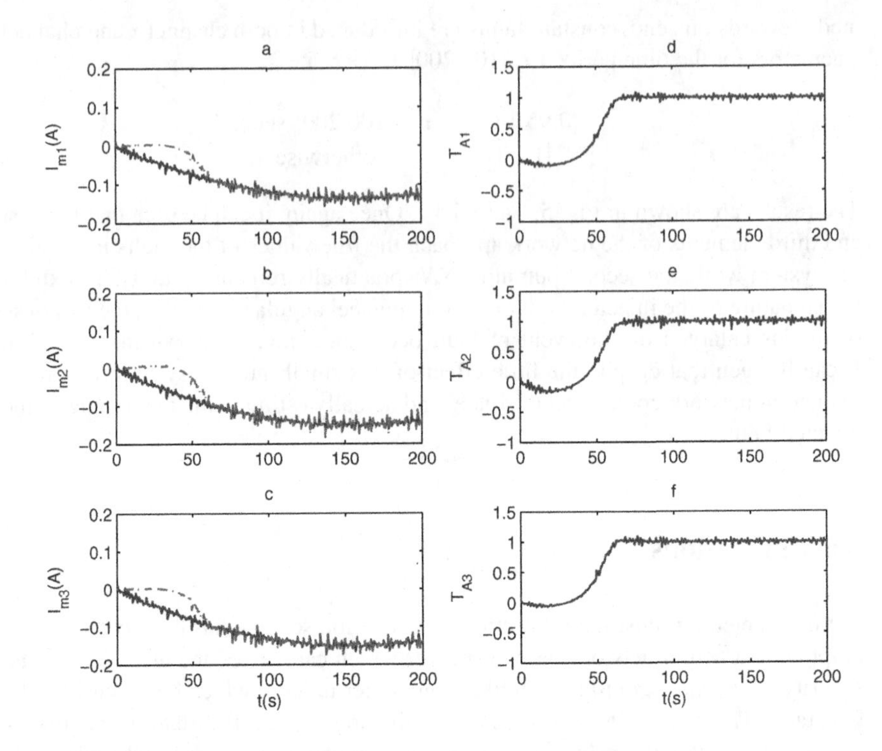

Fig. 5.2 Actuator fault detection system (NN_1) responses to a 0.15 sin(0.01t) reference trajectory in fault-free operation: (a)-(c) estimated and actual motor currents; (d)-(f) estimated actuator gain faults. The solid lines represent the actual wheel motor currents and the dashed lines correspond to the estimated motor currents.

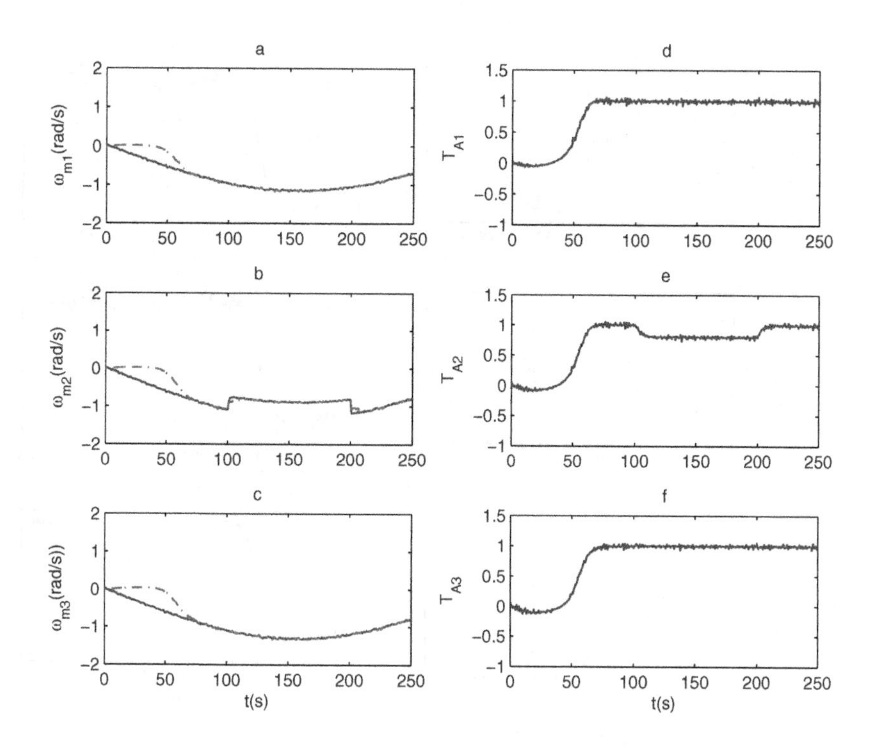

Fig. 5.3 Actuator fault detection and isolation (NN_1) response to a 0.15 sin(0.01t) reference trajectory corresponding to a 20% gain fault in channel y actuator ($T_A = [1\ 0.8\ 0.1]^T$), (a)-(c) estimated and actual wheel angular velocities; (d)-(f) estimated actuator faults. The solid lines represent the actual wheel angular velocities and the dashed lines correspond to the estimated angular velocities.

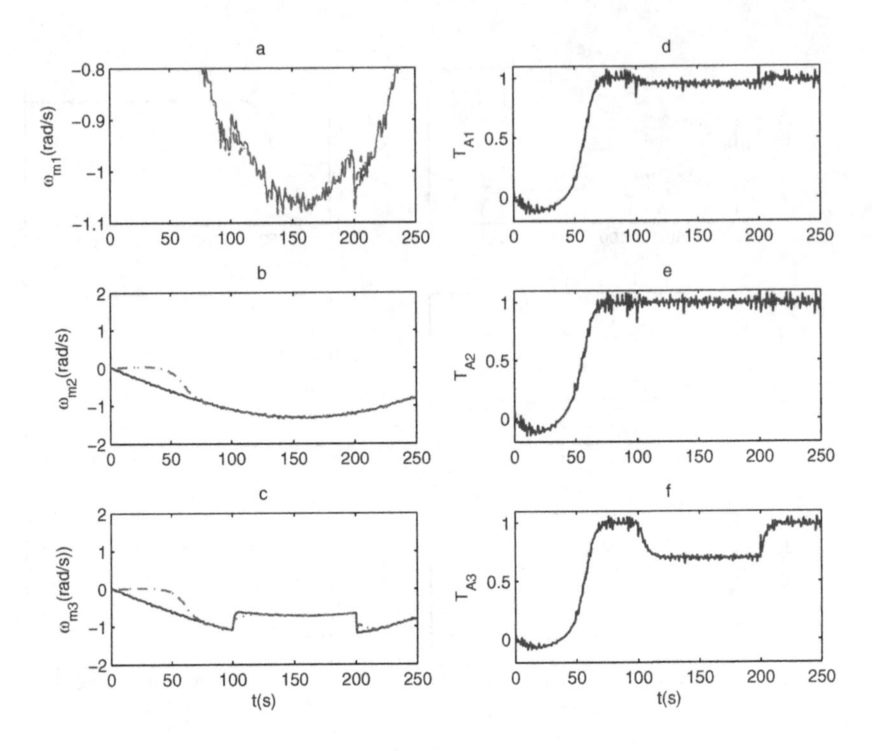

Fig. 5.4 Actuator fault detection and isolation (NN_1) response to a 0.15 sin(0.01t) reference trajectory corresponding to simultaneous faults in the channel x & z actuators ($T_A = [.95\ 1\ 0.7]^T$), (a)-(c) estimated and actual wheel angular velocities; (d)-(f) estimated actuator faults. The solid lines represent the actual wheel angular velocities and the dashed lines correspond to the estimated angular velocities.

Chapter 6
A Robust Sensor and Actuator Fault Detection and Estimation Approach

6.1 Introduction

This chapter presents a robust *actuator* as well as *sensor* fault detection and isolation scheme for a general nonlinear system using a neural network-based observer. The nonlinear system is subject to state and sensor uncertainties and disturbances.

Unlike many previous methods reported in the literature (e.g. [115, 116, 117]), the proposed method does not rely on the availability of all system states. The stability of the overall fault detection and isolation approach in the presence of unknown faults as well as plant and sensor uncertainties is shown by using Lyapounov's direct method. This is a challenging problem due to the presence of two nonlinear-in-parameters neural network and the coupling between the two networks' weights as well as the states of the observer. No restrictive assumption is imposed on the system and/or the fault detection scheme in order to simplify the stability analysis. Furthermore, the stability of the neural network weights and estimation errors are all addressed in a unified fashion. Magnetorquer type actuators and Magnetometer type sensors that are commonly utilized in the attitude determination and control of Low-Earth Orbit (LEO) satellites are considered for case studies. The effectiveness of the proposed fault diagnosis strategy is demonstrated through extensive number of simulation studies.

The remainder of the chapter is organized as follows: Section 6.2 introduces the framework and formulates the FDI problem. In Section 6.3, the neural network-based fault detection and isolation scheme is introduced and a detailed stability analysis is provided. Case studies dealing with fault detection and isolation of the satellite's attitude control subsystem with magnetorquer type actuators and magnetometer type sensor are presented in Section 6.4. Simulation results are shown in Section 6.5, and Section 6.6 concludes the chapter.

H.A. Talebi et al., *Neural Network-Based State Estimation of Nonlinear Systems*, Lecture Notes in Control and Information Sciences 395,
DOI 10.1007/978-1-4419-1438-5_6,

6.2 Problem Statement

Consider the nonlinear system

$$\begin{aligned}\dot{x}(t) &= f(x(t),u(t)) + \eta_x(x,u,t) + T_A(x(t),u(t)) \\ y(t) &= Cx(t) + \eta_y(x,u,t) + T_S(x(t),u(t)),\end{aligned} \tag{6.1}$$

where $u \in \Re^{mu}$ is the input, $y \in \Re^{my}$ is the output, $x \in \Re^n$ is the state vector, $f : \Re^n \times \Re^m \longrightarrow \Re^n$ is the vector-valued known nonlinear function, $\eta_x : \Re^n \times \Re^m \times \Re \longrightarrow \Re^n$ represents the plant unmodeled dynamics and disturbances, $\eta_y : \Re^n \times \Re^m \times \Re \longrightarrow \Re^m$ is the sensor modeling uncertainties and noise, and $T_A : \Re^n \times \Re^m \longrightarrow \Re^n$ is the unknown actuator fault and $T_S : \Re^n \times \Re^m \longrightarrow \Re^m$ represents the unknown sensor fault.

By adding and subtracting Ax, where A is a Hurwitz matrix, one may rewrite (6.1) as

$$\begin{aligned}\dot{x} &= Ax + g(x,u) + \eta_x(x,u,t) + T_A(x,u) \\ y &= Cx + \eta_y(x,u,t) + T_S(x,u),\end{aligned} \tag{6.2}$$

where $g(x,u) = f(x,u) - Ax$. Note that for the sake of simplicity the time dependency of the system variables is not explicitly shown in the above equation.

To facilitate the design and analysis of our proposed fault detection and isolation (FDI) scheme the following assumptions are made:

Assumption 6.1 *The nonlinear system (6.2) is observable.*

Assumption 6.2 *The nominal closed-loop system is stable, i.e., there exists a control function $u = k(y) = k(Cx)$ such that $\dot{x} = f_k(Cx) = f(x,k(Cx))$ is asymptotically stable. An immediate conclusion is that the linearized Jacobian J_0 defined below*

$$J_0 = \left.\frac{\partial f_k}{\partial x}\right|_{x=0},$$

is Hurwitz.

Assumption 6.3 *The plant and sensor uncertainty vectors η_x and η_y are uniformly bounded, i.e.,*

$$\begin{aligned}\|\eta_x\| &\le \bar{\eta}_x \\ \|\eta_y\| &\le \bar{\eta}_y.\end{aligned} \tag{6.3}$$

Assumption 6.4 *The function $g(x)$ is Lipschitz in x with Lipschitz constant l_g, i.e.,*

$$\|g(x,u) - g(\hat{x},u)\| \le l_g \|\tilde{x}\|, \tag{6.4}$$

where $\tilde{x} = x - \hat{x}$.

Assumption 6.5 *It is assumed that the sensor and actuator faults do not occur simultaneously. In other words, both vectors $T_A(x,u)$ and $T_S(x,u)$ cannot be nonzero at any given time.*

Therefore, the fault detection and isolation problem can be formally stated as follows:

Objectives of the FDI problem: Given Assumptions 6.1 to 6.5, develop a methodology that can detect, isolate, and estimate the fault vectors T_A and T_S in a stable fashion.

It is worth noting that, by using the proposed FDI scheme, if the fault vectors are estimated correctly, the isolation and detection objectives are satisfied simultaneously. Indeed, the neural network-based methodology presenting in the next section provides the fault detection, isolation, and estimation in a unified framework. In other words, to isolate the actuator/sensor faults neither additional measured/calculated signals nor a separate fault isolation policy is required.

6.3 Neural Network-Based Fault Detection and Estimation Scheme for Sensor/Actuator Faults

Considering the representation (6.2), recurrent models are provided by parameterizing the maps T_A and T_S through employing feedforward neural network architectures, denoted by NN_1 and NN_2, respectively. Therefore, the following model is considered for our observer-based FDI scheme

$$\begin{aligned}\dot{\hat{x}} &= A\hat{x} + g(\hat{x},u) + \hat{T}_A(\hat{x},u,W_1),\\ \hat{y} &= C\hat{x} + \hat{T}_S(\hat{x},u,W_2),\end{aligned} \tag{6.5}$$

where $\hat{T}_A(\hat{x},u,W_1)$ and $\hat{T}_S(\hat{x},u,W_2)$ are the maps performed by the neural networks NN_1 and NN_2, respectively. The structure of the estimators is shown in Fig. 6.1.

It is a well-known result (see Theorem 1.1) that for x restricted to a compact set S of $x \in \Re^n$ and for some sufficiently large number of hidden layer neurons, there exist weights and thresholds such that any continuous function on the compact set S can be represented by a recurrent NLPNN as

$$\begin{aligned}T_A(x,u) &= W_1\sigma(V_1\bar{x}) + \varepsilon_1(\bar{x})\\ T_S(x,u) &= W_2\sigma(V_2\bar{x}) + \varepsilon_2(\bar{x}),\end{aligned}$$

where $\bar{x} = \begin{bmatrix} x \\ u \end{bmatrix}$, W_1, V_1, W_2, and V_2 are the neural networks ideal weight matrices, $\varepsilon_1(\bar{x}) \le \varepsilon_N$ and $\varepsilon_2(\bar{x}) \le \varepsilon_N$ are the neural network bounded approximation error, and $\sigma(.)$ is the transfer function of the hidden neurons which is assumed to be a tangent hyperbolic function given in (2.4).

The functions $T_A(x,u,w)$ and $T_S(x,u,w)$ may be approximated as

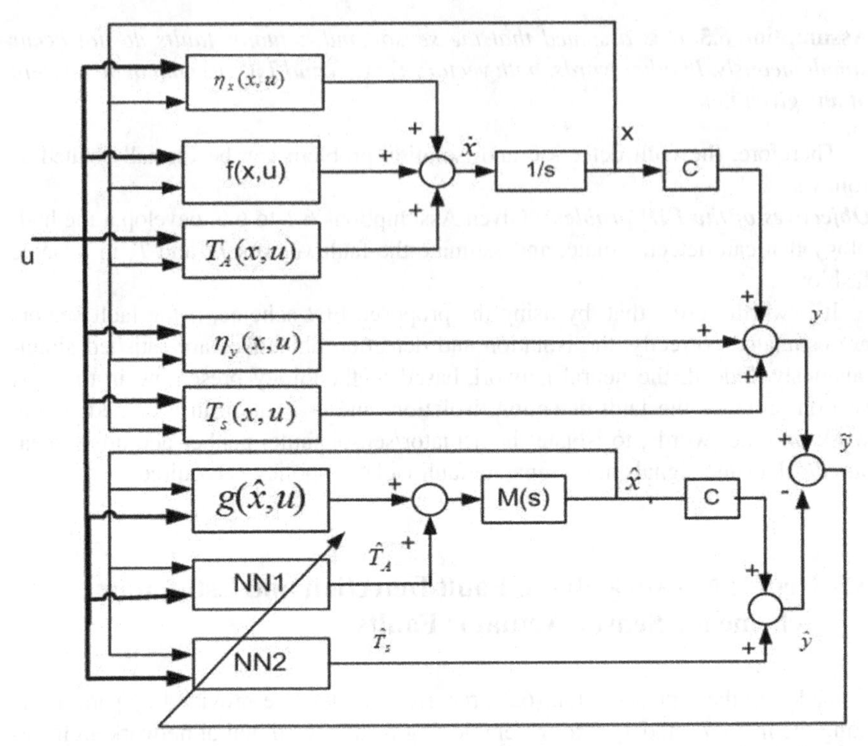

Fig. 6.1 Structure of the neural network-based fault detection and isolation scheme for sensor/actuator faults.

$$\hat{T}_A(\hat{\bar{x}}, W_1) = \hat{W}_1 \sigma(\hat{V}_1 \hat{\bar{x}}) \tag{6.6}$$

$$\hat{T}_S(\hat{\bar{x}}, W_2) = \hat{W}_2 \sigma(\hat{V}_2 \hat{\bar{x}}), \tag{6.7}$$

where $\hat{\bar{x}} = \begin{bmatrix} \hat{x} \\ u \end{bmatrix}$. Therefore, the estimator model can be described as

$$\begin{aligned} \dot{\hat{x}}(t) &= A\hat{x} + g(\hat{x}, u) + \hat{W}_1 \sigma(\hat{V}_1 \hat{\bar{x}}) \\ \hat{y} &= C\hat{x} + \hat{W}_2 \sigma(\hat{V}_2 \hat{\bar{x}}). \end{aligned} \tag{6.8}$$

Moreover, using (6.2), (6.6) , (6.7), and (6.8), and defining the state estimation error as $\tilde{x} = x - \hat{x}$, and the output estimation error as $\tilde{y} = y - \hat{y}$ lead us to state the error dynamics by the following equations:

$$\begin{aligned} \dot{\tilde{x}}(t) &= A\tilde{x} + g(\bar{x}) - g(\hat{\bar{x}}) + \eta_x(\bar{x}) + \tilde{W}_1 \sigma(\hat{V}_1 \hat{\bar{x}}) + w_1(t) \\ \tilde{y} &= C\tilde{x} + \eta_y(\bar{x}) + \tilde{W}_2 \sigma(\hat{V}_2 \hat{\bar{x}}) + w_2(t), \end{aligned} \tag{6.9}$$

where $w_1(t) = W_1[\sigma(V_1\bar{x}) - \sigma(\hat{V}_1\hat{\bar{x}})] + \varepsilon_1(\bar{x})$ and $w_2(t) = W_2[\sigma(V_2\bar{x}) - \sigma(\hat{V}_2\hat{\bar{x}})] + \varepsilon_2(\bar{x})$ are bounded disturbance terms [80], i.e., $\|w_1(t)\| \leq \bar{w}_1$ and $\|w_2(t)\| \leq \bar{w}_2$

for some positive constants $\bar{w}_1$ and $\bar{w}_2$, due to the characteristics of the tangent hyperbolic function.

It can be seen that $\hat{V}_1$ and $\hat{V}_2$ appear in (6.9) only as arguments of the tangent hyperbolic function which is known to be bounded. Now, let us introduce new bounded variables S_1 and S_3 defined as

$$\begin{aligned} S_1 &= \sigma(\hat{V}_1\hat{\bar{x}}) \\ S_3 &= \sigma(\hat{V}_2\hat{\bar{x}}). \end{aligned} \tag{6.10}$$

Therefore, Equation (6.9) can be rewritten as

$$\begin{aligned} \dot{\tilde{x}}(t) &= A\tilde{x} + g(\bar{x}) - g(\hat{\bar{x}}) + \eta_x(\bar{x}) + \tilde{W}_1 S_1 + w_1(t) \\ \tilde{y} &= C\tilde{x} + \eta_y(\bar{x}) + \tilde{W}_2 S_3 + w_2(t), \end{aligned} \tag{6.11}$$

The next section presents an FDI policy for sensor and actuator faults of nonlinear system (6.1).

6.3.1 Fault Detection and Isolation Policy

Similar to [16] and the FID scheme introduced in Chapter 5, the employed neural networks identify a fault and produce small (ideally zero) output in non-faulty (healthy) operation. When there is a fault in the actuator/sensor, the output of the neural network NN_1/NN_2, i.e., $\hat{T}_A/\hat{T}_S$ is nonzero and that is used as a residual to detect and identify the type of the fault. In fact, an actuator fault is declared if the residual $\hat{T}_A$ exceeds a certain threshold value. Similarly, a sensor fault is declared by comparing $\hat{T}_S$ to a given threshold value. Once this threshold is exceeded the presence of a fault is concluded.

6.3.2 System Identification, Fault Detection, and Stability Analysis

Theorem 6.1 (see below) provides the learning rules for updating the weights of the neural networks to guarantee identification and representation of the nonlinear system as well as detection and isolation of fault signatures for system (6.1). The stability analysis of the fault detection and isolation scheme is also shown by utilizing Lyapunov's direct method.

Theorem 6.1. *Consider the nonlinear system (6.1) and the nonlinear observer model (6.5). Given Assumptions 6.1 to 6.5, if the weights of the neural networks NN_1 and NN_2 are updated according to:*

$$\dot{\hat{W}}_1 = -\eta_1(\frac{\partial J}{\partial \hat{W}_1}) - \rho_1\|\tilde{y}\|\hat{W}_1 \quad (6.12)$$

$$\dot{\hat{V}}_1 = -\eta_2(\frac{\partial J}{\partial \hat{V}_1}) - \rho_2\|\tilde{y}\|\hat{V}_1 \quad (6.13)$$

$$\dot{\hat{W}}_2 = -\eta_3(\frac{\partial J}{\partial \hat{W}_2}) - \rho_3\|\tilde{y}\|\hat{W}_2 \quad (6.14)$$

$$\dot{\hat{V}}_2 = -\eta_4(\frac{\partial J}{\partial \hat{V}_2}) - \rho_4\|\tilde{y}\|\hat{V}_2, \quad (6.15)$$

then $\tilde{x}, \tilde{y}, \tilde{W}_1$, $\tilde{V}_1, \tilde{W}_2$, and $\tilde{V}_2 \in L_\infty$, i.e., the state and output estimation errors, and the weights error are all ultimately bounded in the presence of unknown actuator and sensor faults as well as plant and sensor uncertainties and noise.

Remark 6.1 *In the above mentioned equations, η_1 to η_4 are positive constants representing the learning rates, ρ_1 to ρ_4 are some small positive constants, and $J = \frac{1}{2}(\tilde{y}^T\tilde{y})$ is the neural network objective function.*

Proof: The proof of the theorem is divided into three parts: In Part 1, the detailed expression for the learning rule (6.12) to (6.15) and also the rational for decomposing the system into two subsystems are given. Part 2 presents the stability of Subsystem 1, i.e., the ultimate boundedness of the state estimation error $\tilde{x}$ and neural networks output weights errors $\tilde{W}_1$ and $\tilde{W}_2$. The stability of Subsystem 2, i.e., the ultimate boundedness of the neural networks hidden weights errors, $\tilde{V}_1$ and $\tilde{V}_2$ is demonstrated in Part 3.

<u>**Part 1:**</u> First, by using the chain rule and static gradient approximation, the learning rules (6.12)-(6.15) can be obtained as follows (the details of algebraic manipulations leading to these equations are given in Appendix C):

$$\dot{\hat{W}}_1 = -\eta_1(\tilde{y}^T C J_0^{-1})^T S_1^T - \rho_1\|\tilde{y}\|\hat{W}_1 \quad (6.16)$$

$$\dot{\hat{V}}_1 = -\eta_2(\tilde{y}^T C J_0^{-1}\hat{W}_1 S_2)^T \hat{\bar{x}}^T - \rho_2\|\tilde{y}\|\hat{V}_1 \quad (6.17)$$

$$\dot{\hat{W}}_2 = -\eta_3(\tilde{y}^T A_c^{-1})^T S_3^T - \rho_3\|\tilde{y}\|\hat{W}_2 \quad (6.18)$$

$$\dot{\hat{V}}_2 = -\eta_4(\tilde{y}^T A_c^{-1}\hat{W}_2 S_4)^T \hat{\bar{x}}^T - \rho_4\|\tilde{y}\|\hat{V}_2, \quad (6.19)$$

where S_1 and S_3 are defined in (6.10) and S_2 and S_4 are given by

$$S_2 = I - \Lambda(\hat{V}_1\hat{\bar{x}})$$
$$S_4 = I - \Lambda(\hat{V}_2\hat{\bar{x}}), \quad (6.20)$$

and

$$\Lambda(\hat{V}_j\hat{\bar{x}}) = diag\{\sigma_i^2(\hat{V}_j^i\hat{\bar{x}})\}, j = 1,2,\ i = 1,2,...,m, \quad (6.21)$$

and $\hat{V}_j^i$ denotes the ith row of the weight matrix $\hat{V}_j$. Now, recall that $\tilde{W}_1 = W_1 - \hat{W}_1$, $\tilde{W}_2 = W_2 - \hat{W}_2$ and define $\tilde{V}_1 = V_1 - \hat{V}_1$ and $\tilde{V}_2 = V_2 - \hat{V}_2$, where W_1, W_2, V_1, and V_2

are the fixed ideal neural network weights. Hence, the learning rules (6.16)-(6.19) in terms of the weight errors may be expressed as:

$$\dot{\tilde{W}}_1 = l_1 \tilde{y} S_1^T + \rho_1 \| \tilde{y} \| \hat{W}_1 \tag{6.22}$$
$$\dot{\tilde{V}}_1 = S_2^T \hat{W}_1^T l_2 \tilde{y} \hat{\bar{x}}^T + \rho_2 \| \tilde{y} \| \hat{V}_1 \tag{6.23}$$
$$\dot{\tilde{W}}_2 = l_3 \tilde{y} S_3^T + \rho_3 \| \tilde{y} \| \hat{W}_2 \tag{6.24}$$
$$\dot{\tilde{V}}_2 = S_4^T \hat{W}_2^T l_4 \tilde{y} \hat{\bar{x}}^T + \rho_4 \| \tilde{y} \| \hat{V}_2, \tag{6.25}$$

where $l_1 = \eta_1 J_0^{-T} C^T$, $l_2 = \eta_2 J_0^{-T} C^T$, $l_3 = \eta_3 A_c^{-T}$, $l_4 = \eta_4 A_c^{-T}$, and A_c is a Hurwitz matrix selected to control the speed of convergence (see Appendix C).

Clearly, the following observations can be made:

Fact 6.1.1 *The boundedness of $\tilde{V}_1$ and $\tilde{V}_2$ has no effect on the stability of $\tilde{x}$ (6.11), $\tilde{W}_1$ (6.22), and $\tilde{W}_2$ (6.24) since S_1 and $S_3 \in L_\infty$ regardless of the value of $\tilde{V}_1$ and $\tilde{V}_2$.*

Fact 6.1.2 *The function $\Lambda(.)$ defined in (6.21) is bounded due to the boundedness of the sigmoidal function, hence S_2 and $S_4 \in L_\infty$.*

Given Fact 6.1.1, we can conclude that the system can be decomposed into two subsystems for the purpose of stability analysis. One subsystem (Subsystem 1) consists of the estimation error dynamics (6.11) and the output-layer weight errors (6.22) and (6.24). The other subsystem (Subsystem 2) consists of the hidden layer weight errors (6.23) and (6.25). Note that, the two subsystems are not completely decoupled due to the presence of $\tilde{x}$, $\hat{W}_1$, and $\hat{W}_2$ in (6.23) and (6.25). Consequently, the stability of Subsystem 1 can be shown independent of Subsystem 2, but not vice versa.

Part 2: To study the stability of Subsystem 1, let us select the following positive definite Lyapunov function candidate

$$L = \frac{1}{2} \tilde{x}^T P \tilde{x} + \frac{1}{2} tr(\tilde{W}_1^T \rho_1^{-1} \tilde{W}_1) + \frac{1}{2} tr(\tilde{W}_2^T \rho_3^{-1} \tilde{W}_2), \tag{6.26}$$

where $P = P^T > 0$ satisfies the Lyapunov equation:

$$A^T P + PA = -Q, \tag{6.27}$$

for the Hurwitz matrix A and some positive-definite matrix Q. The time derivative of (6.26) is given by

$$\dot{L} = \frac{1}{2} \dot{\tilde{x}}^T P \tilde{x} + \frac{1}{2} \tilde{x}^T P \dot{\tilde{x}} + tr(\tilde{W}_1^T \rho_1^{-1} \dot{\tilde{W}}_1) + tr(\tilde{W}_2^T \rho_3^{-1} \dot{\tilde{W}}_2). \tag{6.28}$$

Substituting (6.11), (6.22), (6.24) and (6.27) into (6.28) leads to

$$\begin{aligned}\dot{L} &= -\frac{1}{2}\tilde{x}^T Q\tilde{x} + \tilde{x}^T P(\tilde{W}_1 S_1 + w_1(t) + g(\bar{x}) - g(\hat{\bar{x}}) + \eta_x(\bar{x})) \\ &+ tr(\tilde{W}_1^T(l_1\rho_1^{-1}\tilde{y}S_1^T) + \tilde{W}_1^T\|\tilde{y}\|(W_1 - \tilde{W}_1)) \\ &+ tr(\tilde{W}_2^T(l_3\rho_3^{-1}\tilde{y}S_3^T) + \tilde{W}_2^T\|\tilde{y}\|(W_2 - \tilde{W}_2)).\end{aligned} \tag{6.29}$$

Denote that for W_{M1}, W_{M2} and σ_m representing the upper bounds for the fixed ideal weights W_1 and W_2 and the tangent hyperbolic function $\sigma(.)$, the following inequalities hold:

$$\begin{aligned} tr(\tilde{W}_1^T(W_1 - \tilde{W}_1)) &\le W_{M1}\|\tilde{W}_1\| - \|\tilde{W}_1\|^2 \\ tr(\tilde{W}_1^T l_1\rho_1^{-1}\tilde{y}S_1^T) &\le \sigma_m\rho_1^{-1}\|\tilde{W}_1\|\ \|l_1\|\ \|\tilde{y}\| \\ tr(\tilde{W}_2^T(W_2 - \tilde{W}_2)) &\le W_{M2}\|\tilde{W}_2\| - \|\tilde{W}_2\|^2 \\ tr(\tilde{W}_2^T l_3\rho_3^{-1}\tilde{y}S_3^T) &\le \sigma_m\rho_3^{-1}\|\tilde{W}_2\|\ \|l_3\|\ \|\tilde{y}\|.\end{aligned} \tag{6.30}$$

Next, (6.11) implies that

$$\|\tilde{y}\| \le \|C\|\|\tilde{x}\| + \bar{\eta}_y + \sigma_m\|\tilde{W}_2\| + \bar{w}_2. \tag{6.31}$$

Consequently, by using (6.3), (6.4), (6.30), and (6.31), we get

$$\begin{aligned}\dot{L} &\le -\frac{1}{2}\lambda_{min}(Q)\|\tilde{x}\|^2 + \|\tilde{x}\|\|P\|\left(\sigma_m\|\tilde{W}_1\| + \bar{w}_1 + l_g\|\tilde{x}\| + \bar{\eta}_x\right) \\ &+ \sigma_m\rho_1^{-1}\|\tilde{W}_1\|\|l_1\|\left(\|C\|\|\tilde{x}\| + \bar{\eta}_y + \sigma_m\|\tilde{W}_2\| + \bar{w}_2\right) \\ &+ \left(\|C\|\|\tilde{x}\| + \bar{\eta}_y + \sigma_m\|\tilde{W}_2\| + \bar{w}_2\right)\left(W_{M1}\|\tilde{W}_1\| - \|\tilde{W}_1\|^2\right) \\ &+ \sigma_m\rho_3^{-1}\|\tilde{W}_2\|\|l_3\|\left(\|C\|\|\tilde{x}\| + \bar{\eta}_y + \sigma_m\|\tilde{W}_2\| + \bar{w}_2\right) \\ &+ \left(\|C\|\|\tilde{x}\| + \bar{\eta}_y + \sigma_m\|\tilde{W}_2\| + \bar{w}_2\right)\left(W_{M2}\|\tilde{W}_2\| - \|\tilde{W}_2\|^2\right) = F,\end{aligned} \tag{6.32}$$

where $\lambda_{min}(Q)$ is the minimum eigenvalue of Q. Now, by rearranging (6.32), we get:

$$\begin{aligned} F &= -\alpha_1\|\tilde{x}\|^2 + \alpha_2\|\tilde{x}\| + \alpha_3\|\tilde{x}\|\|\tilde{W}_1\| + \alpha_4\|\tilde{W}_1\| - \alpha_5\|\tilde{W}_1\|^2\|\tilde{x}\| - \alpha_6\|\tilde{W}_1\|^2 \\ &+ \alpha_7\|\tilde{W}_2\|\|\tilde{x}\| + \alpha_8\|\tilde{W}_2\| - \alpha_9\|\tilde{W}_2\|^2\|\tilde{x}\| - \alpha_{10}\|\tilde{W}_2\|^2 + \alpha_{11}\|\tilde{W}_2\|^2 \\ &- \alpha_{12}\|\tilde{W}_2\|^3 + \alpha_{13}\|\tilde{W}_1\|\|\tilde{W}_2\| - \alpha_{14}\|\tilde{W}_2\|\|\tilde{W}_1\|^2,\end{aligned} \tag{6.33}$$

where

$$
\begin{aligned}
\alpha_1 &= \frac{1}{2}\lambda_{min}(Q) - l_g\|P\|,\\
\alpha_2 &= (\bar{w}_1 + \bar{\eta}_x)\|P\|,\\
\alpha_3 &= \sigma_m(\|P\| + \rho_1^{-1}\|l_1\|\|C\|) + W_{M1}\|C\|,\\
\alpha_4 &= \bar{\eta}_y(\rho_1^{-1}\sigma_m\|l_1\| + W_{M1}) + \bar{w}_2(W_{M1} + \sigma_m\rho_1^{-1}\|l_1\|),\\
\alpha_5 &= \|C\|,\\
\alpha_6 &= \bar{\eta}_y + \bar{w}_2,\\
\alpha_7 &= \sigma_m\rho_3^{-1}\|l_3\|\|C\| + W_{M2}\|C\|,\\
\alpha_8 &= \left(\rho_3^{-1}\sigma_m\|l_3\| + W_{M2}\right)\left(\bar{\eta}_y + \bar{w}_2\right),\\
\alpha_9 &= \|C\|,\\
\alpha_{10} &= \bar{\eta}_y + \bar{w}_2,\\
\alpha_{11} &= \sigma_m W_{M2} + \sigma_m^2\rho_3^{-1}\|l_3\|,\\
\alpha_{12} &= \sigma_m,\\
\alpha_{13} &= \rho_1^{-1}\sigma_m^2\|l_1\| + \sigma_m W_{M1},\\
\alpha_{14} &= \sigma_m.
\end{aligned}
$$

Note that all parameters defined above are positive except α_1 which can be kept positive if $\lambda_{min}(Q) > 2l_g\|P\|$. By completing the squares for the terms involving $\|\tilde{W}_1\|$ and $\|\tilde{W}_2\|$, we try to find some conditions on $\|\tilde{x}\|$, independent of the neural network weights errors that make the derivative of the Lyapunov function candidate negative semi-definite. Now, by performing some algebraic manipulations, the details of which are provided in Appendix D, we obtain the conditions that ensure the negative semi-definiteness of $\dot{L}$. In fact, provided that $\lambda_{min}(Q) > 2l_g\|P\|$, $\dot{L} \le 0$ if

$$
\|\tilde{x}\| \ge x^0 = \frac{(\alpha_2 + \beta_1) + \sqrt{(\alpha_2 + \beta_1)^2 + 4\beta_4\alpha_1}}{2\alpha_1}, \tag{6.34}
$$

Or

$$
\|\tilde{W}_2\| \ge w_2^0 = \frac{\beta_2 + \sqrt{\beta_2^2 + 4\beta_5(\alpha_{10} + \alpha_{12})}}{2(\alpha_{10} + \alpha_{12})}, \tag{6.35}
$$

where β_i, $i = 1, \ldots, 5$ are given in Appendix D.

Equations (6.34) and (6.35) specify the balls χ_x and χ_w according to

$$
\begin{aligned}
\chi_x &= \{\tilde{x} \mid \|\tilde{x}\| > x^0\}\\
\chi_w &= \{\tilde{W}_2 \mid \|\tilde{W}_2\| > w_2^0\}.
\end{aligned}
$$

Indeed, growing $\tilde{x}$ or $\tilde{W}_2$ outside the balls χ_x or χ_w yields negativeness of $\dot{L}$ and hence the reduction of L, which in turn results in the reduction of its argument $\tilde{x}$ and $\tilde{W}_2$. In other words, $\tilde{x}$ and $\tilde{W}_2$ are ultimately bounded with bound of χ_x and χ_w respectively. The ultimate boundedness of $\tilde{y}$ immediately follows from (6.11) since $\tilde{x}$, $\eta_y(x)$, $\tilde{W}_2$, S_3, and w_2 are all bounded. To show the ultimate boundedness of $\tilde{W}_1$, consider (6.22) which can be expressed as

$$\begin{aligned}\dot{\tilde{W}}_1 &= f_1(\tilde{y}) + \rho_1\|\tilde{y}\|\hat{W}_1 \\ &= f_1(\tilde{y}) + \gamma_1 W_1 - \gamma_1\tilde{W}_1,\end{aligned} \tag{6.36}$$

where

$$\begin{aligned}f_1(\tilde{y}) &= l_1\tilde{y}S_1^T \\ \gamma_1 &= \rho_1\|\tilde{y}\|.\end{aligned}$$

It can be seen that $f_1(.)$ is bounded since $\tilde{y}$ and S_1 are both bounded, C is bounded, and J_0 is a Hurwitz matrix. Given the fact that the ideal weight W_1 is fixed, (6.36) can be regarded as a linear system with bounded input $(f_1(\tilde{y}) + \gamma_1 W_1)$. It is clear that this system is stable since γ_1 is positive and the system input remains bounded. Hence, the ultimate boundedness of $\tilde{W}_1$ is also ensured.

Part 3: The above analysis proves the stability of Subsystem 1, i.e., the estimation error and neural network output weights errors. In the sequel, the proof of stability of Subsystem 2 is given. Recall that Subsystem 2 consists of neural networks hidden layer weights error dynamics (6.23) and (6.25) which can be rearranged as

$$\begin{aligned}\dot{\tilde{V}}_1 &= f_2(\tilde{y},\hat{W}_1,\hat{\bar{x}}) + \rho_2\|\tilde{y}\|\hat{V}_1 \\ &= f_2(\tilde{y},\hat{W}_1,\hat{\bar{x}}) + \gamma_2 V_1 - \gamma_2\tilde{V}_1\end{aligned} \tag{6.37}$$

$$\begin{aligned}\dot{\tilde{V}}_2 &= f_3(\tilde{y},\hat{W}_2,\hat{\bar{x}}) + \rho_4\|\tilde{y}\|\hat{V}_2 \\ &= f_2(\tilde{y},\hat{W}_2,\hat{\bar{x}}) + \gamma_3 V_2 - \gamma_3\tilde{V}_2,\end{aligned} \tag{6.38}$$

where

$$\begin{aligned}f_2(\tilde{y},\hat{W}_1,\hat{\bar{x}}) &= S_2^T\hat{W}_1^T l_2\tilde{y}\hat{\bar{x}}^T \\ f_3(\tilde{y},\hat{W}_2,\hat{\bar{x}}) &= S_4^T\hat{W}_2^T l_4\tilde{y}\hat{\bar{x}}^T \\ \gamma_2 &= \rho_2\|\tilde{y}\| \\ \gamma_3 &= \rho_4\|\tilde{y}\|.\end{aligned}$$

It can be easily seen from the above equations that $f_2(.)$ and $f_3(.)$ are bounded since all their arguments have been previously shown to be bounded. Hence, similar to the analysis given for $\tilde{W}_1$, (6.37) and (6.38) can be regarded as linear systems with bounded inputs $f_2(\tilde{y}) + \gamma_2 V_1$ and $f_3(\tilde{y}) + \gamma_3 V_2$, respectively. Consequently, the stability of (6.37) and (6.38) (hence, the ultimate boundedness of $\tilde{V}_1$ and $\tilde{V}_2$) is also guaranteed since γ_2 and γ_3 are positive. This completes the proof of the theorem. □

6.4 A Case Study: Application to a Satellite's Attitude Control Subsystem

In the sequel, the fault detection and isolation scheme developed in the previous section is applied to magnetorquer type actuators as well as magnetometer type sensors used in the satellite attitude control subsystem.

6.4.1 Dynamic Modeling

Let us recall the dynamic equation of a satellite attitude model from Section 4.3

$$H\dot{\omega} = -S(\omega)H\omega + \tau,$$

where $\tau \in \Re^3$ is the control input vector, $\omega = [\omega_1\ \omega_2\ \omega_3]^T$ is the angular velocity vector of the satellite, expressed in the satellite body frame, H is the symmetric positive definite inertia matrix of the satellite, and $S(\omega)$ is the cross product matrix given by

$$S(\omega) = \begin{bmatrix} 0 & \omega_3 & -\omega_2 \\ -\omega_3 & 0 & \omega_1 \\ \omega_2 & -\omega_1 & 0 \end{bmatrix}.$$

6.4.2 Magnetorquer Model (ACS Actuator)

A principal diagram of the coil driven electronics and a magnetorquer is shown in Fig. 6.2 [6]. When an electric current is passed through a coil, a magnetic dipole is generated. Such dipoles tends to align themselves with any external magnetic field, and this property is used to produce a torque for the purpose of attitude control. The torque produced is related to the cross product of the magnetic dipole of the coil (**M**) and the local geomagnetic field vector (**B**) according to:

$$\tau = \mathbf{M} \times \mathbf{B} = \begin{bmatrix} M_yB_z - M_zB_y \\ M_zB_x - M_xB_z \\ M_xB_y - M_yB_x \end{bmatrix}. \tag{6.39}$$

The strength of the magnetic dipole depends on the number of coils (N), the current passing through them (I), the cross sectional area that the coils encompass (A) and the permeability of the core, namely:

$$\mathbf{M} = kNIA. \tag{6.40}$$

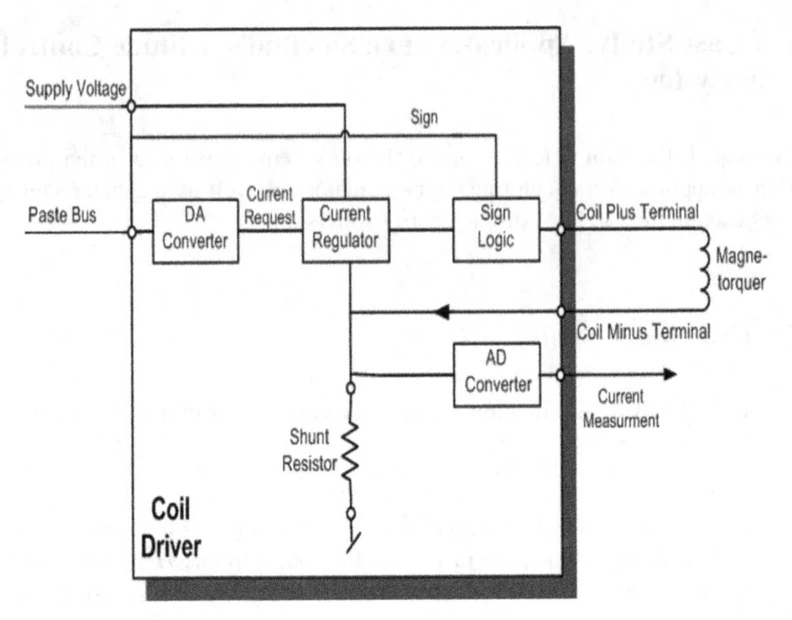

Fig. 6.2 Principal diagram of a single coil driver and magnetorquer [6].

The direction and strength of the geomagnetic field experienced by the satellite depends upon the satellite orbital position. The geomagnetic field is often expressed in a local frame and a transformation due to orbit plane inclination and orientation is required to relate this field. When a current is passed through a coil, two torques are generated depending upon the orientation of the magnetic field as can also be seen from (6.39). For instance, in polar regions the magnetic field vector primarily lies in the yz plane (i.e., $B_x \approx 0$) [128]. Hence, to generate a torque about the z axis (τ_z), an undesirable torque about the y axis would also result (τ_y). This problem known as *cross-torque* is due to the orientation of the magnetic field vector. To limit cross torque, operation of the magnetorquer must be limited to times when the orientation of the magnetic field is favorable.

In summary, an attitude control strategy for operating the torques only when the satellite is within one of the two polar or two equatorial region of the orbit can minimize cross torques. Control about two axes can be achieved in any control region, while the third axis is uncontrolled and under the direct influence of the cross torque. After the satellite has passed through a control region, no active attitude control is used until the satellite enters the next control region. When the satellite is moving between control regions, disturbance torques cause the satellite attitude to drift. Roll and pitch disturbances are limited to oscillations by the gravity boom. Disturbances about the yaw axis are not controlled by the gravity boom. The most common control strategy employed for satellites with magnetorquer is the so-called cross product control rule [6, 128] that is given by:

$$\mathbf{M} = \frac{\tau_d \times \mathbf{B}}{\|\mathbf{B}\|},$$

where τ_d is the desired torque obtained from the attitude and/or rate errors.

6.4.3 Actuator (Magnetorquer) Fault

To express the satellite and actuator dynamics in standard state space representation, we first write (6.39) in a matrix form and then substitute it in (4.10), that is:

$$H\dot{\omega} = -S(\omega)H\omega + \begin{bmatrix} 0 & B_z(t) & -B_y(t) \\ -B_z(t) & 0 & B_x(t) \\ B_y(t) & -B_x(t) & 0 \end{bmatrix} \mathbf{M}. \tag{6.41}$$

Now, by defining $x = \omega$ and $u = \mathbf{M}$, equation (6.41) can be written as

$$\begin{aligned} \dot{x} &= f(x,u), \\ f(x,u) &= H^{-1}(B(t)u - S(\omega)H\omega), \\ B(t) &= \begin{bmatrix} 0 & B_z(t) & -B_y(t) \\ -B_z(t) & 0 & B_x(t) \\ B_y(t) & -B_x(t) & 0 \end{bmatrix}. \end{aligned} \tag{6.42}$$

As seen above, the system is actually a time-varying system since the magnetic field matrix $B(t)$ is time-varying. In [109], the time invariant counterpart of system (6.42) was used for control design. This was done by averaging the components of $B(t)$ over one orbit [129], that is:

$$B = \frac{1}{T}\int_0^T B(t)dt. \tag{6.43}$$

Remark 6.2 *Note that the above approximation results in unmodeled dynamics that can be added to (6.42) as an uncertainty term η_x, that is:*

$$\eta_x = H^{-1}(B(t) - B)u. \tag{6.44}$$

Magnetorquers are subject to various faults, namely low magnetic moment ($M = 0$), low supply voltage, coil wire and/or shunt resistor disconnection and short circuit are just a few to name. A comprehensive list of cause and effects of magnetorquer's fault may be found in [6]. In this chapter, various types of magnetorquer faults may be considered by adding an extra torque component τ_a to the dynamic equations (6.42). Moreover, the output vector of the system is selected as: $y = [\omega_1\ \omega_2]$ resulting in $C = \begin{bmatrix} 1 & 0 & 0 \\ 0 & 1 & 0 \end{bmatrix}$.

$$\dot{x} = f(x,u) + \eta_x + T_A, \qquad (6.45)$$
$$T_A = H^{-1}\tau_a.$$

Hence, when $\tau_a \equiv 0$, the actuator is indeed healthy and a nonzero τ_a would correspond to any of the above sources of anomalies and fault causes.

6.4.4 Magnetometer Model (ACS Sensor)

Spacecraft attitude is the rotational coordinate transformation from a chosen reference system into the body frame. It can be determined from the simultaneous observation of two non-parallel vectors. Typical vector measurements are the body frame direction to the Sun, selected stars, the Earth center, or along the Earth's magnetic field. The true Sun, start, Earth, and geomagnetic field vectors in the reference frame are known from catalog files or analytical models [130].

Magnetometers for detecting the Earth's field are standard equipment on many spacecraft. They are often part of the momentum management system, working in parallel with the onboard Attitude Control System (ACS). Measurement of the geomagnetic field provides a reliable and economical basis for determining the attitude of low-Earth orbiting three-axis stabilized spacecraft. When used in combination with strap-down gyroscope for rate determination, magnetometer data can be filtered to obtain attitude estimates. Moreover, momentum management is part of the control system concerned with dumping excess angular momentum through the use of magnetorquers or thrusters. Spacecraft with magnetorquers use magnetometers to determine the local geomagnetic field to know which combination of torques to activate.

A magnetometer measured signal B_m can be represented by the following model [131]:

$$B_m = B + EB + \Delta B + \nu, \qquad (6.46)$$

where $B_m = [B_{mx}\ B_{my}\ B_{mz}]^T$ is the measured magnetic field, $B = [B_x\ B_y\ B_z]^T$ is the true magnetic field, $\Delta B = [\Delta B_x\ \Delta B_y\ \Delta B_z]^T$ is the magnetometer bias, ν is the measurement noise, and E is the magnetometer misalignment matrix:

$$E = \begin{bmatrix} 0 & E_z & -E_y \\ -E_z & 0 & E_x \\ E_y & -E_x & 0 \end{bmatrix},$$

with $E = [E_x\ E_y\ E_z]^T$ being the vector of small misalignment along each of the three axes. It can be seen that the expression (6.46) is in the form of (6.2). Hence, any fault, misalignment, or noise in the magnetometer measurement can directly contribute to the error in rate measurements, either as noise and uncertainty vector η_y or as an additive fault vector T_S introduced in (6.2).

6.5 Simulation Results

To verify the performance of our proposed fault diagnosis scheme, simulations are performed on the *Ørsted* satellite and the results are presented in this section. The Ørsted satellite was launched by a Delta II launch vehicle on March 1997 into a $450 \times 850km$ orbit with a 96 degree inclination. Four basic sensors (Magnetometer, Star Imager (SIM), Global Positioning System (GPS), and Sun sensor) on board the satellite were used for attitude determination. As for actuators, magnetic torquing was achieved using three perpendicular sets of redundant electromagnetic air cored coils.

The inertia matrix and the other parameters used in the simulations are adopted from [109], namely:

$$H = \begin{bmatrix} 181.78 & 0 & 0 \\ 0 & 181.25 & 0 \\ 0 & 0 & 1.28 \end{bmatrix} kg.m^2.$$

The magnetic field vector in the Ørsted Orbit is obtained by using a 10th order spherical harmonic model [132]. As mentioned earlier, the unmodeled dynamics η_x resulting from approximation (6.43) is also included in the simulation model (6.45). A conventional feedback controller is used to stabilize the closed-loop system corresponding to the fault-free operation.

All sensory measurements are assumed to be corrupted with 10% Gaussian noise. The neural networks approximating actuator and sensor faults, i.e., NN_1 and NN_2, respectively have three layers where the input layer has 6 neurons, the hidden layer has 5 neurons with sigmoidal activation functions, and the output layer has 3 neurons with linear transfer functions. The input to both networks is $\hat{x}$ as defined in Section 6.3 and is shown in Fig. 6.1. The output of NN_1 and NN_2 are $\hat{T}_A = [\hat{T}_{A1}\ \hat{T}_{A2}\ \hat{T}_{A3}\ ]^T$ and $\hat{T}_S = [\hat{T}_{S1}\ \hat{T}_{S2}]^T$, respectively. The numerical values used for the neural networks learning parameters as specified in equations (6.12) to (6.15) are: $\eta_1 = \eta_2 = 20; \eta_3 = \eta_4 = 0.1; \rho_1 = \rho_2 = \rho_3 = \rho_4 = 1e-6$, and the Hurwitz matrix A is selected as $A = -2I_3$, where I_3 denotes a 3×3 identity matrix.

At first, a non-faulty situation was considered and a $0.1sin(0.001t)$ reference trajectory was applied to the satellite. The simulation results for the actual and estimated angular velocities are shown in Fig. 6.3. The output of the neural networks are shown in Fig. 6.4 for NN_1 and NN_2. As can be seen, despite the noisy and corrupted sensory measurements, the magnitudes of the neural network outputs remain close to zero confirming that there is no fault in the system.

Remark 6.3 *Note that the outputs of the networks do not approach to zero due to presence of measurement noise, uncertainties in the satellite, magnetorquer, and magnetometer dynamic models, such as η_x given in (6.44). Therefore, care should be exercised in selecting appropriate values for the fault detection thresholds. Several approaches are proposed for proper selection of a threshold, among which adaptive and fuzzy thresholds are common. For our simulations, several tests were performed under different operating conditions, i.e., different reference trajectories.*

Then, based on the outcome of the tests and the upper bounds assumed for uncertainties and measurement noises (see (6.3) and (6.44)), a threshold value of 0.02 *was selected for both actuator and sensor faults. Simulation results confirm that using this threshold value results in* no false or missing alarm *under different faults and/or reference trajectories in terms of shape and magnitude.*

The fault detection policy is stated as follows: Let t_f be the time when a residual exceeds its corresponding threshold value. An actuator/sensor fault is declared if

$$\begin{array}{c} |\hat{T}_{Ai}| > th_{ai},\ i = 1, \ldots 3, \\ \text{Or} \\ |\hat{T}_{Si}| > th_{si},\ i = 1, \ldots 3 \\ \forall t \in [t_f, t_f + t_r], \end{array}$$

where th_{ai} and th_{si} are the threshold values corresponding to the ith actuator/sensor, respectively. Furthermore, in the above expression t_r is chosen to allow the outputs of neural networks reach their steady state responses. In our simulations, t_r is selected as $t_r = 30$ *second*. Note that in all simulation results, the bounds specified by the above threshold are shown as dotted lines. However, this threshold bound cannot be seen for the cases where the magnitude of the residuals are much smaller than the thresholds.

Next, a 0.1 bias fault was introduced exclusively for channel x magnetorquer for the time period $t \in [1000\ 1500]$ *second*, i.e.,

$$T_A = \begin{cases} [0.1\ 0\ 0]^T & t \in [1000\ 1500] \text{ second} \\ [0\ 0\ 0]^T & \text{otherwise} \end{cases}.$$

The responses of the actual and estimated angular velocities as well as estimated actuator and sensor faults are shown in Fig. 6.5 and Fig. 6.6, respectively. Jumps in angular velocity responses for $t \in [1000\ 1500]$ *second* (see Fig. 6.5) demonstrate the effect of the actuator fault. From Fig. 6.6, one may observe that the first output of NN_1 approaches to a nonzero value corresponding very closely to the actual magnitude of the fault that was injected to only the channel x magnetorquer. It can also be observed that the neural network in fact was able to correctly isolate the location of the inserted fault with an excellent accuracy. Note that the network responses for channels y and z, that is T_{A2} and T_{A3} are practically zero. Furthermore, note that there was no false positive flag produced by the output of NN_2 corresponding to sensor faults.

In the next set of simulation results, the sensitivity and the capability of our proposed fault detection and isolation scheme for identifying simultaneous faults were evaluated. Towards this end, constant faults are introduced in both the x and y channel (axis) actuators for the duration of $t \in [1000\ 1500]$ *second*, i.e.,

$$T_A = \begin{cases} [0.03\ 0.2\ 0]^T & t \in [1000\ 1500] \text{ second} \\ [0\ 0\ 0]^T & \text{otherwise} \end{cases}.$$

The results shown in Fig. 6.7 verify that the first and second outputs of NN_1 converge to true values of the faults inserted in the system. Furthermore, as in the other cases, the isolation of the actuator faults were identified correctly and no false positive flags were generated for the sensor fault T_S.

Sensor fault detection and isolation was considered in the next set of simulation studies. A constant sensor bias fault ($T_S = [0.03\ 0]$) was introduced in the sensor measuring ω_1 after $t = 10$ *second* and the results are shown in Fig. 6.8 and Fig. 6.9. Similar to the case of actuator fault, Fig. 6.8 shows the effect of the sensor fault on satellite angular velocity (ω_1) at $t = 10$ *second*. The results demonstrate that despite noisy measurements, the network NN_2 is able to estimate the true value of the fault T_S, as shown in Fig. 6.9. Again, as in the actuator fault isolation case, we can clearly observe that the network NN_2 responses for output T_{S2} is zero, demonstrating excellent sensor fault isolation.

Next, the same set of simulations as in actuator faults are performed to demonstrate sensor fault separability. Specifically, constant sensor bias faults are introduced in the sensors measuring ω_1 and ω_2 after $t = 10$ *second*, i.e., $T_S = [0.03\ 0.05]^T$ and the results are shown in Fig. 6.10. It can be observed that the first and second output of NN_2, i.e., $\hat{T}_{S1}$ and $\hat{T}_{S2}$ approach to the true value of inserted faults while all components of NN_1 never leave the bounds specified by the thresholds.

Then, an attempt is made to evaluate the capability of the proposed scheme to detect, isolate, and identify incipient faults. Towards this end, an incipient fault and and abrupt fault were introduced in the x and y channel (axis) actuators for the duration of $t \in [1000\ 1500]$ *second*, respectively:

$$T_A = \begin{cases} \left[0.1e^{-0.01(t-1000)} \quad 0.1\ 0\right]^T & t \in [1000\ 1500] \text{ second} \\ [0\ 0\ 0]^T & \text{otherwise} \end{cases}. \tag{6.47}$$

The responses of the neural networks and the actual faults inserted in the system are shown in Fig. 6.11. The dash-dotted lines represent the actual faults. As can be observed, the incipient fault in channel x actuator slowly builds up after 1000 *second* and the neural network output T_{A1} closely follow the actual fault from the beginning. Moreover, the abrupt fault in channel y actuator was also correctly identified by the neural network output T_{A2}.

Finally, another set of simulation was performed to evaluate the performance of the FDI scheme under different operating condition. For this purpose, a reference trajectory with much higher magnitude, namely $0.5sin(0.001t)$ was applied to the satellite with a very small fault was injected exclusively for channel y magnetorquer for the time period $t \in [500\ 1500]$ *second*, i.e.,

$$T_A = \begin{cases} [0\ 0.05\ 0]^T & t \in [500\ 1500] \text{ second} \\ [0\ 0\ 0]^T & \text{otherwise} \end{cases}.$$

The responses of the fault detection and isolation system are shown in Fig. 6.12. It can be observed that **only** the second element of the neural network output, i.e., T_{A2} approaches to a nonzero value indicating there is a fault in channel y magnetorquer confirming the excellent isolation property of the proposed FDI approach. To exhibit the signature of the injected fault on satellite angular velocity, the response of ω_2 is enlarged on the event of fault occupance, i.e., for $t \in [480\ 530]$ *second* as shown in Fig. 6.12-b. It can be seen that despite the little effect of this small fault on system responses, the neural network could detect, isolate, and actually estimate the magnitude of the injected fault. Note that except for T_{A2}, the other components of neural networks never leave the bounds specified by the thresholds.

As mentioned in Remark 6.2, the magnetorquer model used for simulations *do include* state uncertainty terms due to the approximation 6.43 and this to some extent shows the robustness of the proposed FDI scheme. To further investigate this issue, the following state and sensor uncertainty terms have been included in the model and another set of simulations have been performed.

$$\eta_x(x,u) = \begin{bmatrix} \varepsilon_1 sin(\omega_2\ \tau_1) \\ \varepsilon_1 sin(\omega_1\ \tau_3) \\ 0 \end{bmatrix}$$
$$\eta_y(x,u) = \begin{bmatrix} sin(\omega_1 e^{-2\omega_2}) \\ sin(\omega_2 e^{-2\omega_3}) \end{bmatrix}. \tag{6.48}$$

First, a fault-free condition was considered and the response of the FDI scheme was obtained by applying a $0.1sin(0.001t)$ reference trajectory to the satellite, for $\varepsilon_1 = 0.02$ and $\varepsilon_2 = 0.05$. The simulation results are shown in Fig. 6.13. As can be seen, despite the presence of the state and sensor uncertainties as well as measurement noise the magnitudes of the neural network outputs remain in the bound specified by the thresholds.

Next, simultaneous faults in channel x and y magnetorquers (i.e., Equation (6.47)) were injected and the simulation was repeated. The simulation results are shown in Fig. 6.14-a – 6.14-e, for T_A and T_S. It can be observed that **only** the first and second element of NN1 neural network, i.e., T_{A1} and T_{A2} approach to a nonzero value confirming the fault in channel x and y magnetorquers while T_{A3} and both elements of T_S stay close to zero. The performance of the sensor fault detection in the presence of the same uncertainties were examined in the next simulation. Simultaneous faults (i.e. $T_S = [0.03\ 0.03]^T$) in both sensors were injected. The simulation results are shown in Fig. 6.15. The results confirm that NN2 outputs ($\hat{T}_S$) approach to their corresponding faults whereas those of NN1 ($\hat{T}_A$) remain below the threshold.

The above scenarios were repeated by different values of ε_1 and ε_2 and **no false alarm** or missed detection was observed for $\varepsilon_1 \leq 0.07$ or $\varepsilon_2 \leq 0.2$. The responses of the proposed FDI scheme for $\varepsilon_1 = 0.08$ and $\varepsilon_2 = 0.1$ in *fault-free* condition are shown in Fig. 6.16. It can be seen that despite the fact that neural network can estimate the uncertainties, the second output of NN1 (i.e., $\hat{T}_{A2}$) exceeds the threshold since the magnitude of the uncertainties are much larger than the bounds using which the thresholds were selected, i.e., $\bar{\eta}_x$ and $\bar{\eta}_y$.

6.6 Conclusions

A robust Fault Detection and Isolation (FDI) scheme for a general nonlinear system subject to state and sensor uncertainties and noise was presented in this chapter. The methodology is based on neural network-based observer strategy and can simultaneously identify both actuator and sensor faults. The stability of the overall fault detection and isolation scheme in the presence of unknown sensor and actuator faults as well as plant and sensor uncertainties and noise was shown by using Lyapunov's direct method with no restrictive assumptions on the system and/or the FDI algorithm. Magnetorquer type actuators and Magnetometer type sensors that are commonly utilized in the attitude determination and control of Low-Earth Orbit (LEO) satellites were considered for the purpose of our case studies. Simulation results provided do indeed demonstrate the effectiveness of the proposed fault diagnosis strategy.

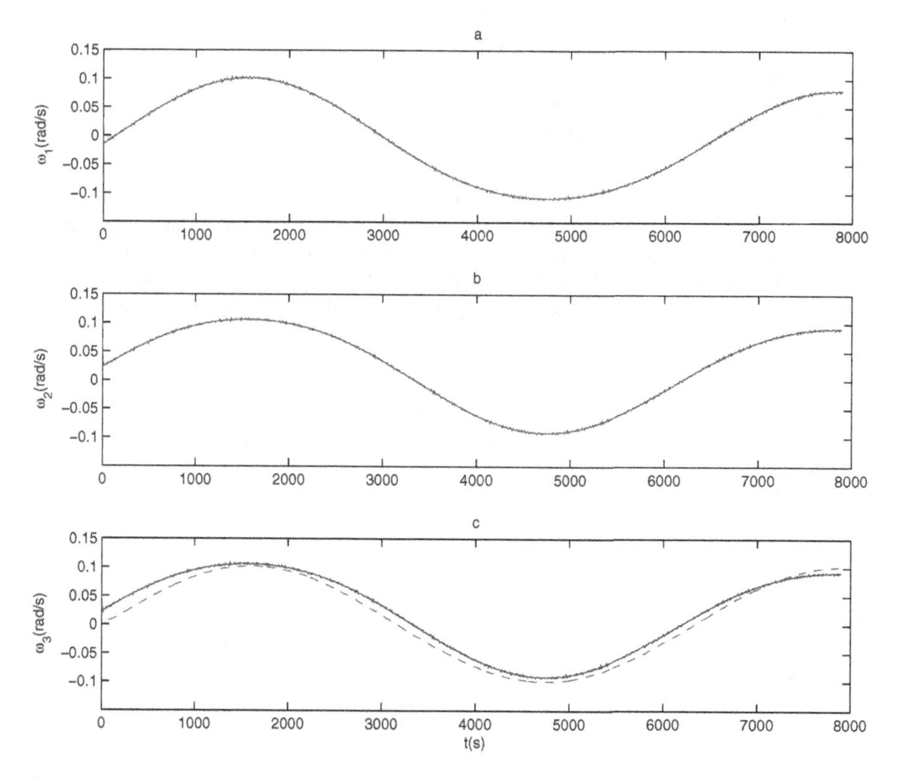

Fig. 6.3 Satellite actual and estimated angular velocities (ω_1 to ω_3) to a $0.1sin(0.001t)$ reference trajectory in a fault-free operation. The solid lines and the dashed lines correspond to the actual and the estimated angular velocities, respectively.

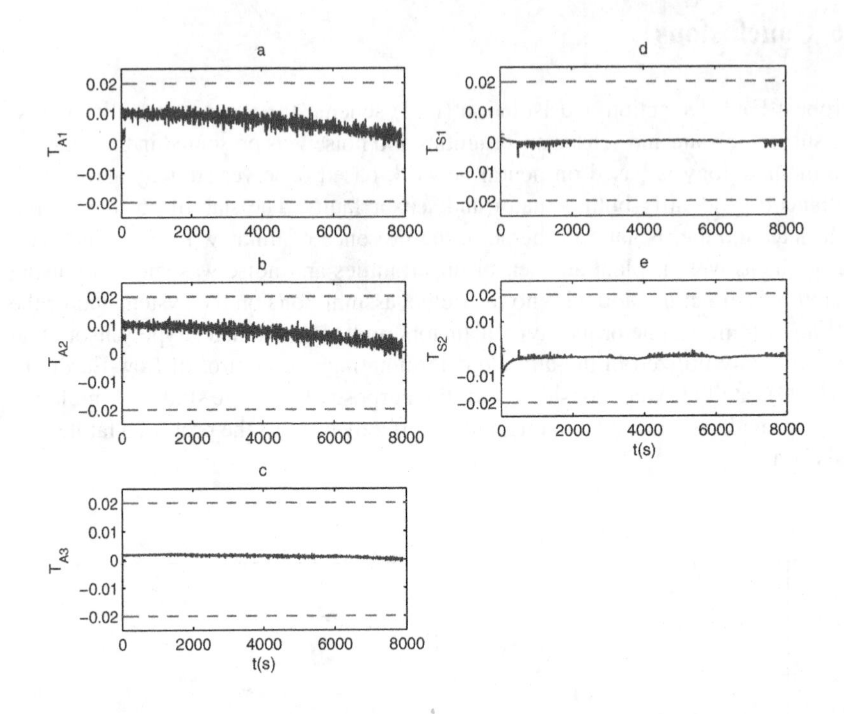

Fig. 6.4 The neural networks NN_1 & NN_2 responses for actuator and sensor fault detection due to $0.1sin(0.001t)$ reference trajectory in a fault-free operation. The thresholds for all sensor and actuator channels are set to $[-0.02\ \ 0.02]$. (a)-(c) estimated actuator faults; (d)-(f) estimated sensor faults. The solid lines represent estimated faults and the dashed lines correspond to the thresholds.

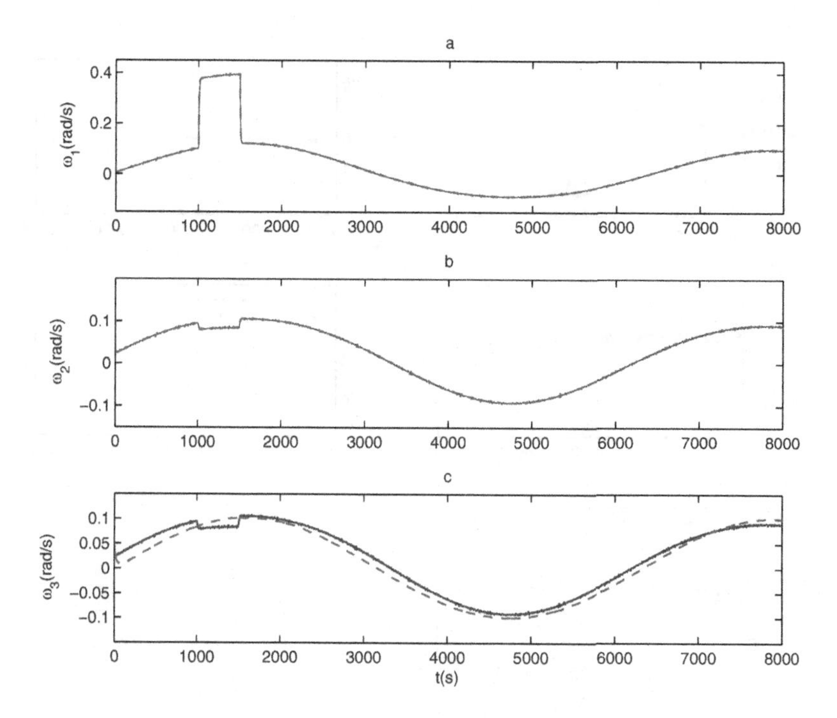

Fig. 6.5 Satellite actual and estimated angular velocities (ω_1 to ω_3) to $0.1sin(0.001t)$ reference trajectory with a bias fault in the channel x magnetorquer ($T_A = [0.1\ 0\ 0]^T$). The solid and dashed lines correspond to the actual and the estimated angular velocities, respectively.

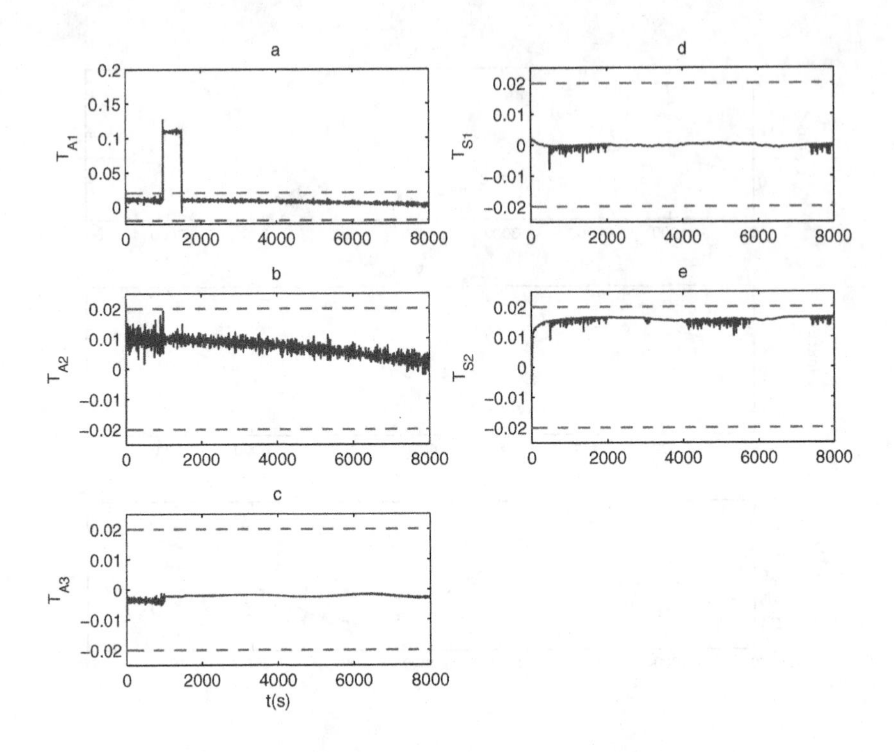

Fig. 6.6 Actuator fault detection and isolation (NN_1) responses to a $0.1sin(0.001t)$ reference trajectory with a constant bias fault in the channel x magnetorquer ($T_A = [0.1\ 0\ 0]^T$), (a)-(c) estimated actuator faults; (d)-(f) estimated sensor faults. The solid lines represent estimated faults and the dashed lines correspond to the thresholds. The thresholds for all sensor and actuator channels are set to $[-0.02\ \ 0.02]$.

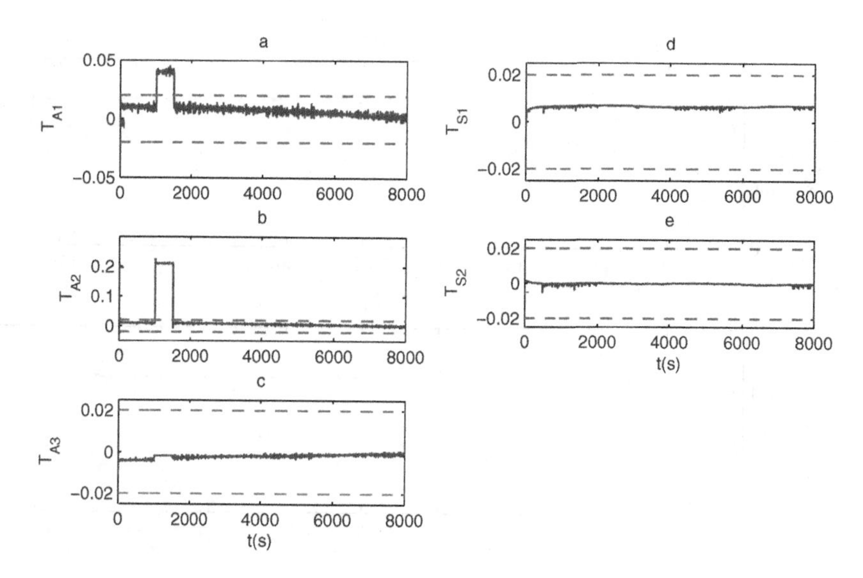

Fig. 6.7 Actuator fault detection and isolation (NN_1) responses to a $0.1sin(0.001t)$ reference trajectory corresponding to simultaneous faults in the channel x & y actuators ($T_A = [0.03\ 0.2\ 0]^T$), (a)-(c) estimated actuator faults; (d)-(f) estimated sensor faults. The solid lines represent estimated faults and the dashed lines correspond to the thresholds.

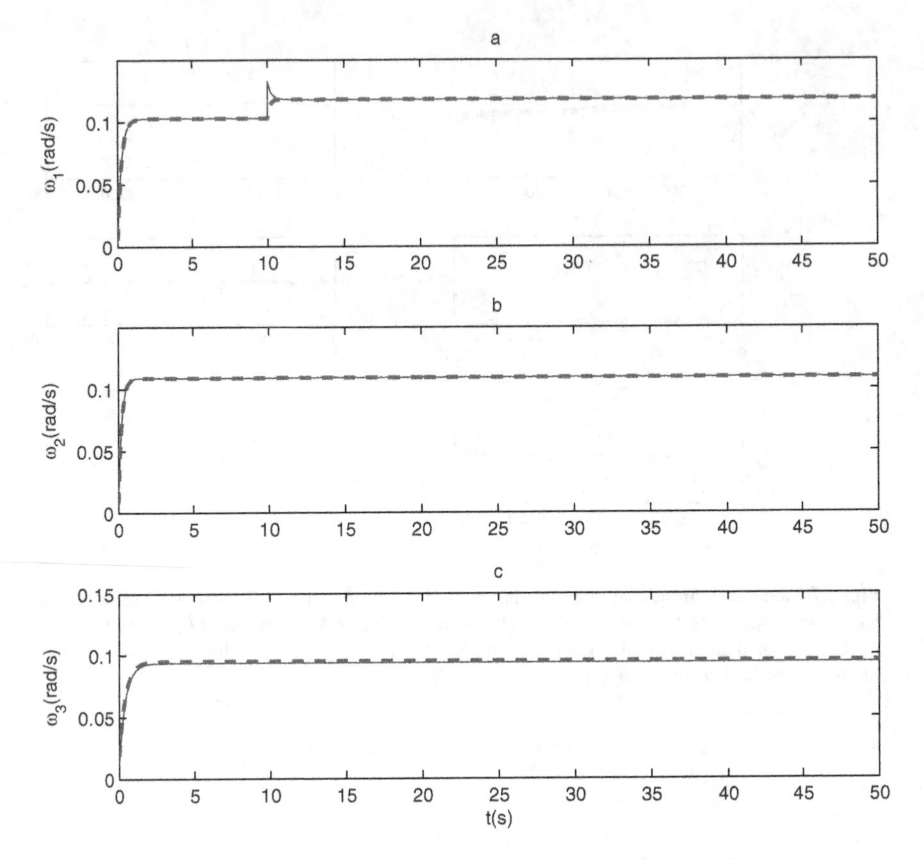

Fig. 6.8 Satellite actual and estimated angular velocities (ω_1 to ω_3) to a step reference trajectory (0.1 rad/s) with a 0.03 rad/s constant fault in ω_1 measurement ($T_S = [0.03\ 0]^T$). The solid and dashed lines correspond to the actual and the estimated angular velocities, respectively.

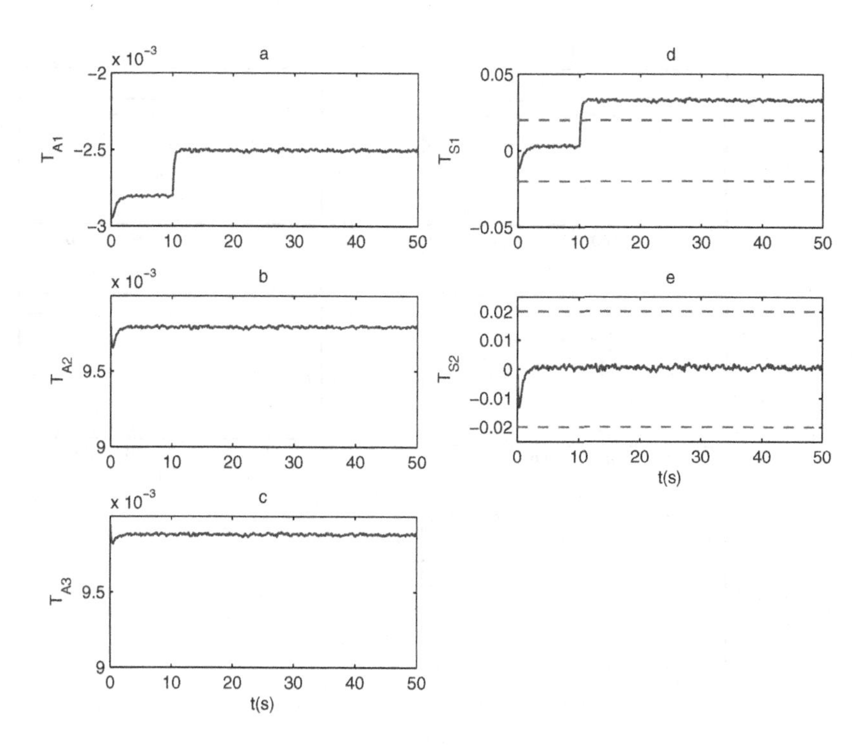

Fig. 6.9 Sensor fault detection and isolation (NN_2) responses to a step reference trajectory (0.1 rad/s) with a 0.03 rad/s constant fault in ω_1 measurement ($T_S = [0.03\ 0]^T$), (a)-(c) estimated actuator faults; (d)-(f) estimated sensor faults. The solid lines represent estimated faults and the dashed lines correspond to the thresholds.

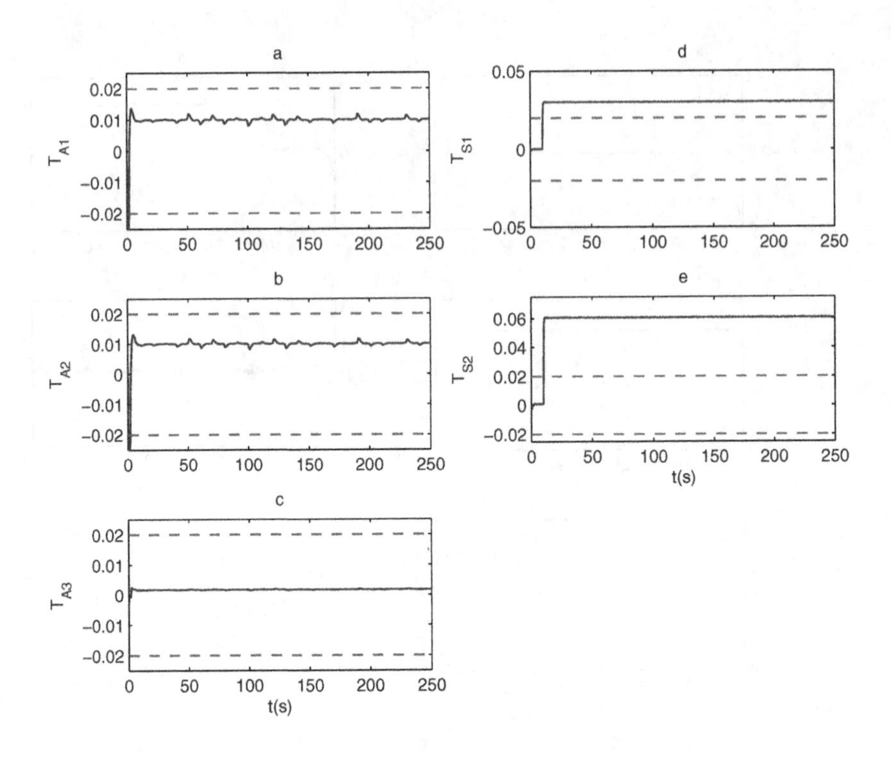

Fig. 6.10 Sensor fault detection and isolation (NN_2) responses to a step reference trajectory (0.1 rad/s) corresponding to simultaneous faults in ω_1 and ω_2 measurements ($T_s = [0.03\ 0.05]^T$), (a)-(c) estimated actuator faults; (d)-(f) estimated sensor faults. The solid lines represent estimated faults and the dashed lines correspond to the thresholds.

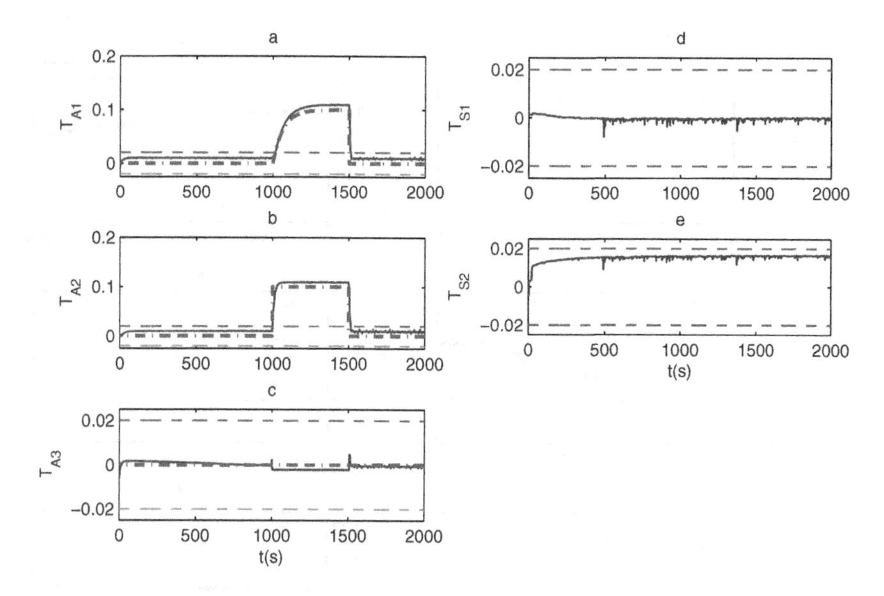

Fig. 6.11 Actuator fault detection and isolation (NN_1) responses to a $0.1sin(0.001t)$ reference trajectory with corresponding to *incipient* and *abrupt* faults in the channel x & y actuators ($T_A = [0.1e^{-0.01(t-1000)}\ 0.1\ 0]^T$) for $t \in [1000,\ 1500]$, (a)-(c) estimated actuator faults; (d)-(f) estimated sensor faults. The solid lines represent estimated faults, the dash-dotted lines represent the actual faults, and the dashed lines correspond to the thresholds.

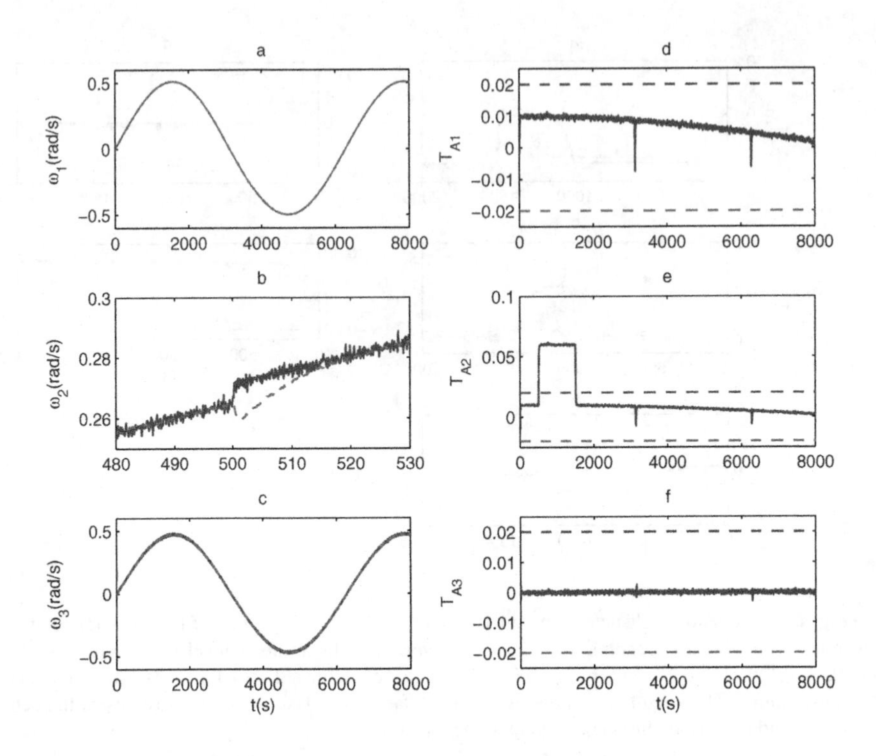

Fig. 6.12 Actuator fault detection and isolation (NN_1) responses to a $0.5sin(0.001t)$ reference trajectory with a small bias fault in the channel y magnetorquer, (a)-(c) angular velocities in x, y and z axes; (d)-(f) estimated faults. The solid lines correspond to the actual angular velocities and the dashed lines correspond to their estimates.

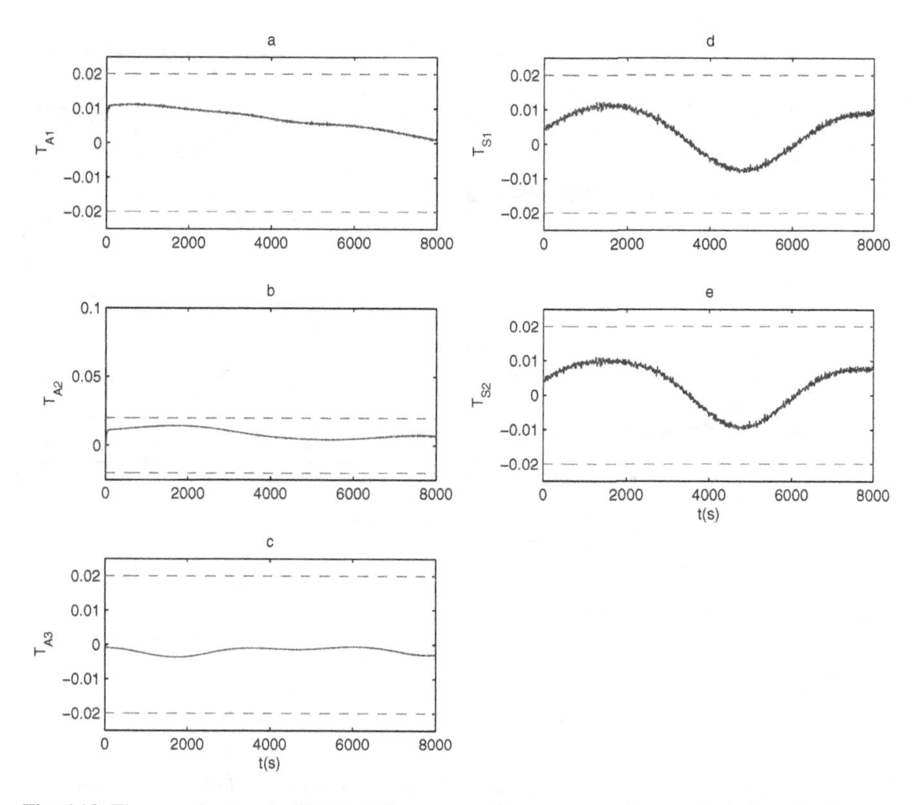

Fig. 6.13 The neural networks NN_1 & NN_2 responses for actuator and sensor fault detection due to $0.1sin(0.001t)$ reference trajectory in a fault-free operation and in the presence of state and sensor uncertainties (6.48). (a)-(c) estimated actuator faults; (d)-(f) estimated sensor faults. The solid lines represent estimated faults and the dashed lines correspond to the thresholds.

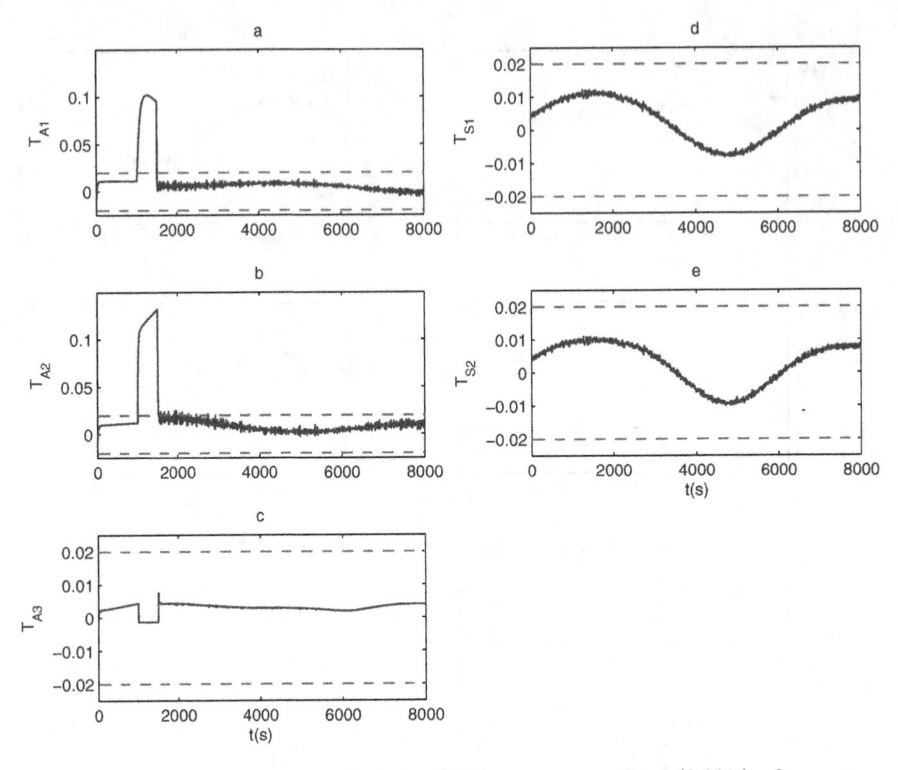

Fig. 6.14 Actuator fault detection and isolation (NN_1) responses to a $0.1sin(0.001t)$ reference trajectory in the presence of state and sensor uncertainties (6.48). (a)-(c) estimated actuator faults; (d)-(f) estimated sensor faults. The solid lines represent estimated faults and the dashed lines correspond to the thresholds.

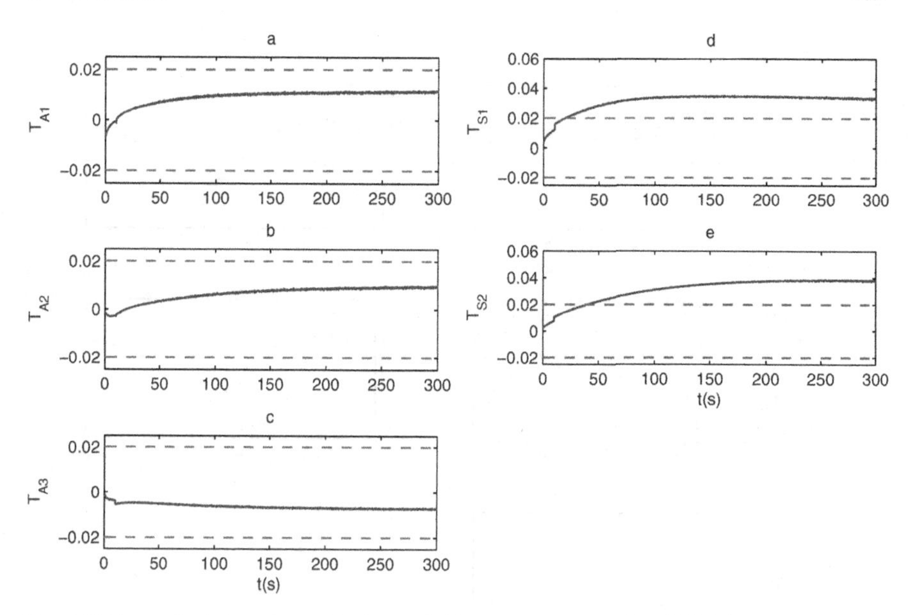

Fig. 6.15 Sensor fault detection and isolation (NN_1) responses to a $0.1sin(0.001t)$ reference trajectory in the presence of state and sensor uncertainties (6.48). (a)-(c) estimated actuator faults; (d)-(f) estimated sensor faults. The solid lines represent estimated faults and the dashed lines correspond to the thresholds.

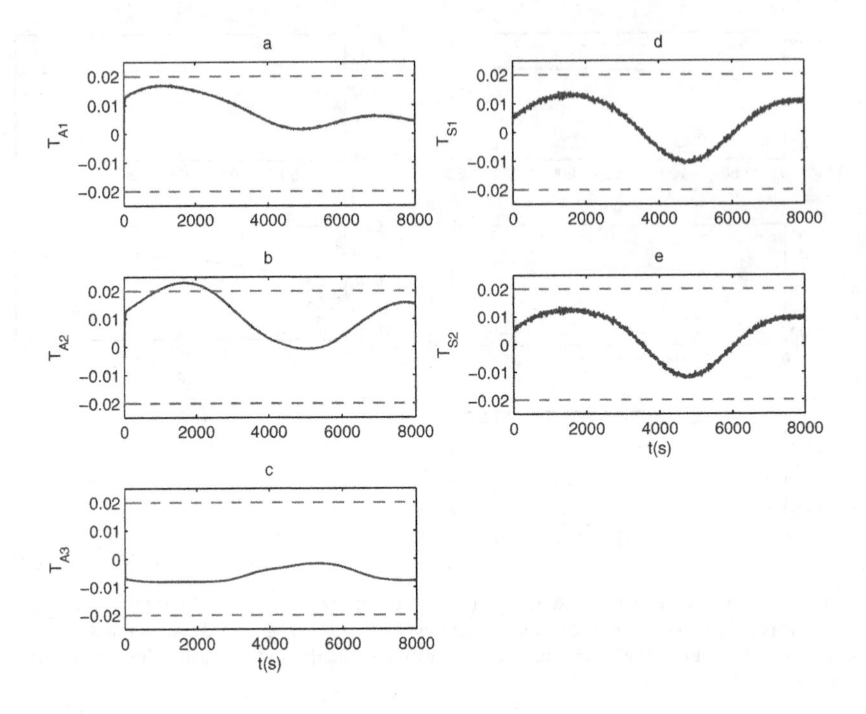

Fig. 6.16 The neural networks NN_1 & NN_2 responses for actuator and sensor fault detection due to $0.1 sin(0.001t)$ reference trajectory in a fault-free operation and in the presence of **large** state and sensor uncertainties (6.48). (a)-(c) estimated actuator faults; (d)-(f) estimated sensor faults. The solid lines represent estimated faults and the dashed lines correspond to the thresholds.

Appendix A
Preliminary Definitions

The goal of this appendix is to briefly explain the preliminary definitions and concepts which are used throughout this monograph.

A.1 Norms

The norm of a vector $x \in R^n$ and the spectral norm of a matrix $A \in R^{m \times n}$ are denoted as

$$||x|| = \sqrt{x^T x}\,, \; ||A||_s = \sqrt{\lambda_{max}[A^T A]},$$

where $\lambda_{max}[.]$ denotes the largest eigenvalue of the positive-definite or positive-semidefinite matrix $[.]$. We denote the smallest eigenvalue of a positive-definite matrix $[.]$ by $\lambda_{\min}[.]$. Given $A = [a_{ij}]$ and $B \in R^{m \times n}$, the Frobenius norm is defined by

$$||A||_F^2 = tr(A^T A) = \Sigma a_{ij}^2,$$

where $tr(.)$ denotes the trace of $(.)$. The associated inner product is

$$< A, B >_F = tr(A^T B).$$

The space of an L_∞ bounded signal can be defined as

$$x(t) \in L_\infty \;\; if \; ess \sup_t |x(t)| < \infty.$$

Now, let $x \in L_\infty$, the L_∞ norm of the signal $x(t)$ is defined as

$$||x||_{L_\infty} = ess \sup_t |x(t)|.$$

H.A. Talebi et al., *Neural Network-Based State Estimation of Nonlinear Systems*,
Lecture Notes in Control and Information Sciences 395,
DOI 10.1007/978-1-4419-1438-5,

A.2 Ultimate Boundedness [1]

Consider the following nonlinear system

$$\dot{x}(t) = f(t,x), \tag{A.1}$$

where $f : [0,\infty) \times D \rightarrow \mathscr{R}^n$ is piecewise continuous in t and locally Lipschitz in x on $[0,\infty) \times D$, and $D \subset \mathscr{R}^n$ is a domain containing the origin. The solution of (A.1) is uniformly ultimately bounded with ultimate bound b, if there exist positive constants b and c, independent of $t_0 \geq 0$, and for every $a \in (0,c)$, there is $T = T(a,b) \geq 0$, independent of t_0, such that

$$\|x(t_0)\| \leq a \Rightarrow \|x(t)\| \leq b, \forall t \geq t_0 + T.$$

A.3 Positive Real and Strictly Positive Real [2]

The concept of Positive Real (PR) and Strictly Positive Real (SPR) transfer function which is derived from network theory, plays important role in the stability of a class of nonlinear systems such as adaptive systems. In [2], the following definitions are given for PR and SPR transfer functions:

Definition A.1. A rational $G(s)$ of the complex variable $s = \sigma + j\omega$ is called **PR** if

- $G(s)$ is real for real s.
- $\mathscr{R}[G(s)] \geq 0 \ \forall \mathscr{R}[s] > 0$

Definition A.2. Assume that $G(s)$ is not identically zero for all s. Then $G(s)$ is **SPR** if $G(s-\varepsilon)$ is PR for some $\varepsilon > 0$.

Appendix B
Flexible-Joint Manipulator Model

In designing robots, developing lighter manipulators which are capable of handling heavier payloads can be achieved by using suitable gears, since by employing a well designed gear, smaller actuators can deliver larger torques. However, one of the disadvantages of adding gears is the presence of "backlash" which reduces the effective bandwidth of the controller since high frequency content in the control signal produces noisy and destructive operation.

Harmonic drives are a special type of transmission providing high torque, high ratio gearing and has been shown to be virtually "backlash" free. One of the most commonly known disadvantages of harmonic drives, however, is the existence of structural flexibility.

In the rest of this appendix, the structure of harmonic derive and its operating principal is presented in Section B.1. Then, in Section B.2 the dynamic model of manipulators are given and the effect of joint flexibility imposed to the system by utilizing harmonic derive is mathematically modeled.

B.1 Harmonic Drive

The harmonic drive which is indeed an input/output gearing mechanism was first introduced by C.W. Musser in 1957 [133]. Then, it was successfully utilized by Hasegawa Gear Works, Ltd. and USM Co., Ltd in 1964. Harmonic drive is applied for gearing reduction, increasing rotational speed or for differential gearing. The harmonic drive theory is based on elastic dynamics and applied the flexibility of metal. As shown in Fig. B.1, the mechanism consists of three basic components:

- **Wave generator** which contains an outer ball bearing and a steel disk called a wave generator plug. A specially designed ball bearing pressed around the bearing plug makes the bearing conform the same elliptical shape as the wave generator plug. Wave generator is usually attached to a servo motor and considered as input member.

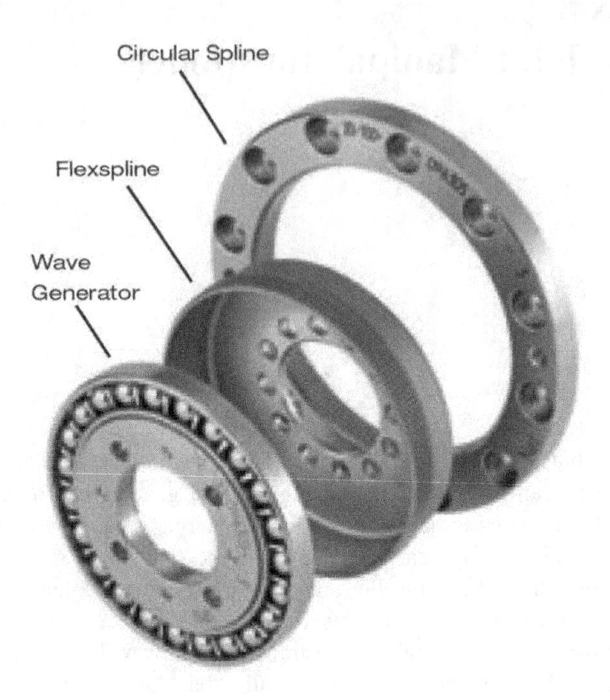

Fig. B.1 Components of harmonic drive [7]

- **Flex spline** which is like a shallow cup, is thin on the sides, but thick and rigid at the bottom. The thin wall results in significant flexibility of the walls at the open end but quite rigid and tightly secured in the closed side. Teeth are designed radially around the outside of the flex spline. The flex spline fits tightly over the wave generator, such that when the wave generator plug is rotating, the flex spline deforms to the shape of a rotating ellipse but does not rotate with the wave generator. The flex spline is considered as the outer member of the mechanism.
- **Circular spline** that is a rigid circular steel ring including teeth on the inside. It is usually attached to the housing and does not rotate. The flex spline and wave generator are inside the circular spline and the teeth of the circular spline mesh with the teeth of the flex spline. Since the flex spline has an elliptical shape, its teeth only mesh with the teeth of the circular spline in two regions on opposite sides of the flex spline, along the major axis of the ellipse.

More complex versions may have a fourth component applied to shorten the overall length or to increase the gear reduction within a smaller diameter. But they still follow the same basic principles.

The trick point of the harmonic drive is that, the flex spline has two fewer teeth than the circular spline. Therefore, when the wave generator rotates one revolution,

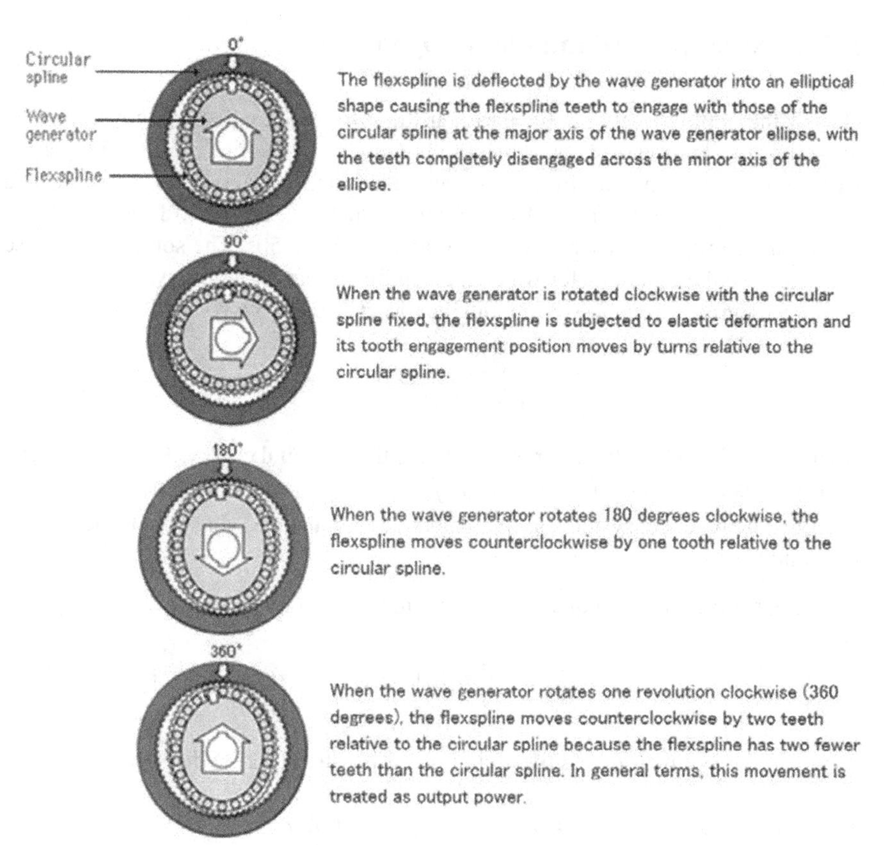

Fig. B.2 Principal operation of harmonic drive [8]

the flex spline and circular spline shift by two teeth. Fig. B.2 depict the principal of harmonic drive operation in details.

Due to the flexible structure of harmonic drive, the mechanical systems using harmonic drive for transmissions suffer from flexibility in joints which makes controlling the systems more challenging. In the following section, the modeling process of joint flexibility in manipulators is discussed.

B.2 Flexible-joint Manipulator Dynamics

One of the popular methods for deriving the differential equations governing the motion of mechanical systems subject to holonomic constrains is **Euler-Lagrange equations** of motion. The Euler-Lagrange equation, or Lagrange's equation, is a differential equation developed by Swiss mathematician Leonhard Euler and Italo-French mathematician Joseph Louis Lagrange in the 1750s. The solutions of these equations are the functions for which a given functional is stationary.

Euler-Lagrange equations of motion are described as follows [134, 3]

$$\frac{d}{dt}\frac{\partial L}{\partial \dot{q}_j} - \frac{\partial L}{\partial q_j} = \tau_j, \tag{B.1}$$

where $L = KE - VE$ is the Lagrangian, VE is the potential energy, KE is the kinetic energy, q_j is coordinate of jth particle, $\tau_j = \psi_j + \frac{\partial VE}{\partial q_j}$, and ψ_j is generalized force.

For some practical systems such as robotic manipulators which holds the following conditions:

- The kinetic energy is a quadratic function of the vector $\dot{q}$, i.e.,

$$KE = \frac{1}{2}\sum_{i,j}^{n} d_{ij}(q)\dot{q}_i\dot{q}_j = \frac{1}{2}D(q)\dot{q}, \tag{B.2}$$

  where the symmetric, positive definite $n \times n$ matrix $D(q)$ is the so-called inertia matrix.
- The potential energy $VE = V(q)$ is independent of $\dot{q}$.

the Euler-Lagrange equations can be derived as follows:

$$L = KE - VE = \frac{1}{2}\sum_{i,j}^{n} d_{ij}(q)\dot{q}_i\dot{q}_j - V(q).$$

Therefore, one can get

$$\begin{aligned}
\frac{\partial L}{\partial \dot{q}_k} &= \sum_j d_{kj}(q)\dot{q}_j \\
\frac{d}{dt}\frac{\partial L}{\partial \dot{q}_j} &= \sum_j d_{kj}(q)\ddot{q}_j + \sum_{i,j}\frac{\partial d_{kj}}{\partial q_i}\dot{q}_i\dot{q}_j \\
\frac{\partial L}{\partial q_k} &= \frac{1}{2}\sum_{i,j}\frac{\partial d_{ij}}{\partial q_k}\dot{q}_i\dot{q}_j - \frac{\partial V}{\partial q_k}.
\end{aligned} \tag{B.3}$$

By considering (B.3), interchanging the order of summation, taking advantage of symmetry, and following some algebraic manipulations, the Euler-Lagrange equations (B.1) can be written as

$$\sum_j d_{kj}(q)\ddot{q}_j + \sum_{i,j} c_{ijk}(q)\dot{q}_i\dot{q}_j + \phi_k(q) = \tau_k, \quad k = 1, ..., n, \tag{B.4}$$

where

$$c_{ijk} = \frac{1}{2}\sum_{i,j} \frac{\partial d_{kj}}{\partial q_i} + \frac{\partial d_{ki}}{\partial q_j} - \frac{\partial d_{ij}}{\partial q_k}. \tag{B.5}$$

is known as **Christoffel symbol**, and $\phi_k = \frac{\partial V}{\partial q_k}$. The equation (B.4) can also be written in matrix form as follows:

$$D(q)\ddot{q} + C(q,\dot{q})\dot{q} + g(q) = \tau, \tag{B.6}$$

where the (k, j)th element of the matrix $C(q,\dot{q})$ is given by $c_{kj} = \sum_{i=1}^{n} c_{ijk}(q)\dot{q}_i$. For more details, please refer to [3].

B.2.1 Dynamic Model of a Two-Link Planar Manipulator [3]

In this section, the detail of manipulator model, i.e., Equation (B.6) for a two-link planar manipulator is provided. Towards this end, consider a planar manipulator with two revolute joints as shown in Fig. B.3.

The kinetic energy of the manipulator is

$$KE = \frac{1}{2}\dot{q}\{m_1 J_{v1}^T J_{v1} + m_2 J_{v2}^T J_{v2} + I_1 \begin{bmatrix} 1 & 0 \\ 0 & 0 \end{bmatrix} + I_2 \begin{bmatrix} 1 & 1 \\ 1 & 1 \end{bmatrix}\}\dot{q}, \tag{B.7}$$

where

$$J_{v1} = \begin{bmatrix} -l_{c1}\sin(q_1) & 0 \\ l_{c1}\cos(q_1) & 0 \\ 0 & 0 \end{bmatrix}$$
$$J_{v2} = \begin{bmatrix} -l_1\sin(q_1) - l_{c2}\sin(q_1+q_2) & -l_{c2}\sin(q_1+q_2) \\ l_1\cos(q_1) + l_{c2}\cos(q_1+q_2) & l_{c2}\cos(q_1+q_2) \\ 0 & 0 \end{bmatrix},$$

and m_i is the total mass of link i, l_i is the total length of link i, l_{ci} is the distance from the joint axis to the ith link center of mass, and I_i is the inertia of link i. The first two terms of (B.7) are related to translational part and the last two terms represent the rotational part of the kinetic energy. Therefore by considering (B.2), $D(q)$ can be defined as follows:

$$D(q) = m_1 J_{v1}^T J_{v1} + m_2 J_{v2}^T J_{v2} + I_1 \begin{bmatrix} I_1 + I_2 & I_2 \\ I_2 & I_2 \end{bmatrix}.$$

Performing the above multiplications and following some manipulations leads to

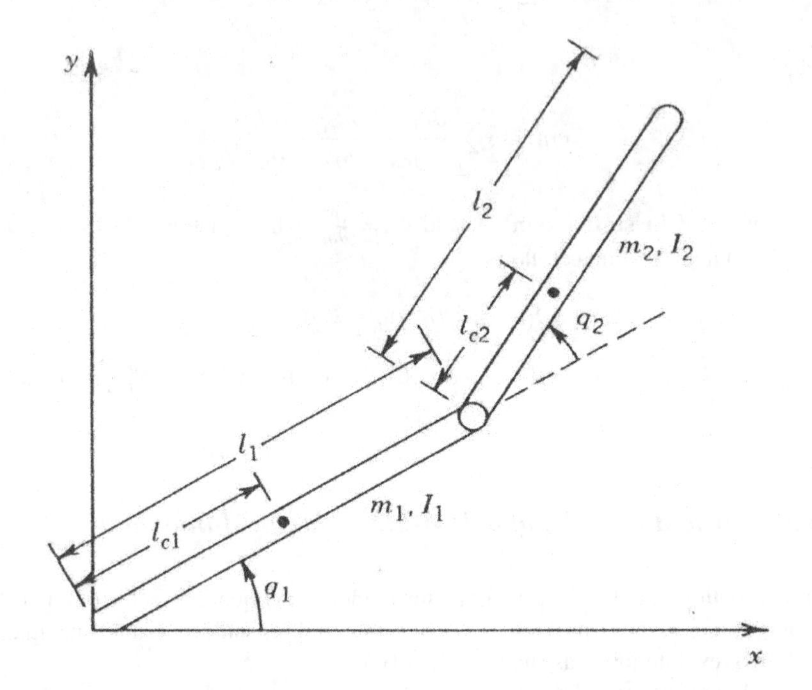

Fig. B.3 Two-link planar manipulator [3].

$$\begin{aligned} d_{11} &= m_1 l_{c1}^2 + m_2(l_1^2 + l_{c2}^2 + 2l_1 l_{c2}\cos(q_2)) + I_1 + I_2 \\ d_{12} &= d_{21} = m_2(l_{c2}^2 + l_1 l_{c2}\cos(q_2)) + I_2 \\ d_{22} &= m_2 l_{c2}^2 + I_2. \end{aligned}$$

Therefore, by using definition (B.5), $C(q,\dot{q})$ can be computed as:

$$C(q,\dot{q}) = \begin{bmatrix} h\dot{q}_2 & h\dot{q}_2 + h\dot{q}_1 \\ -h\dot{q}_1 & 0 \end{bmatrix},$$

where $h = -m_2 l_1 l_{c2}\sin(q_2)$.

On the other hand, the potential energy of the manipulator is given as

$$VE = (m_1 l_{c1} + m_2 l_1) g \sin(q_1) + m_2 l_{c2} g \sin(q_1 + q_2).$$

Hence, $g(q)$ defined in (B.6) is obtained as follows

$$\begin{aligned} g(q) = [\frac{\partial VE}{\partial q_1} \; \frac{\partial VE}{\partial q_2}]^T &= [(m_1 l_{c1} + m_2 l_1) g\cos(q_1) + m_2 l_{c2} g\cos(q_1 + q_2) \\ &= m_2 l_{c2} g\cos(q_1 + q_2)]^T. \end{aligned}$$

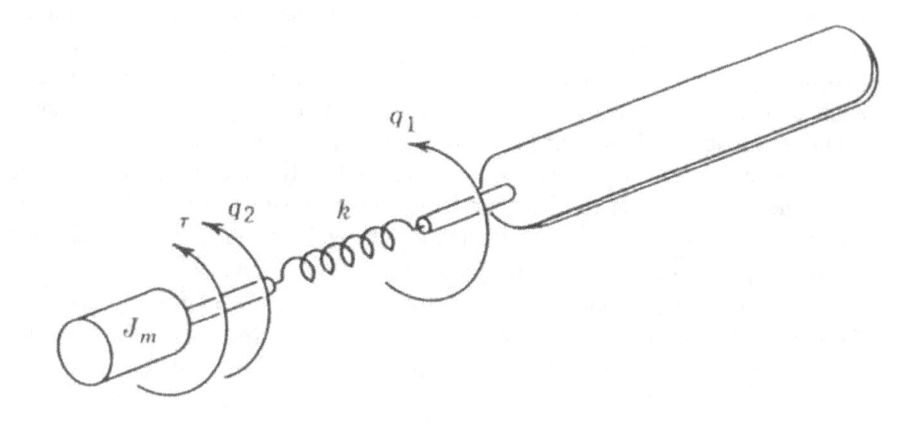

Fig. B.4 Flexible-joint manipulator model [3]

B.2.2 The Effect of Joint Flexibility on Manipulator Model [3]

In this section, the effect of joint flexibility on the dynamic equations of the manipulators is discussed. Generally, such flexibility is described by inserting a rotational spring between the two end of the harmonic drive. It should be noted that due to the presence of the joint flexibility, there are twice as many degree of freedom as in the rigid joint manipulator. For instance the two link manipulator has four states. By refereing to Fig 2.2 which is also shown in Fig. B.4, let us define the coordinates of the system as q_1 corresponding to the link position and q_2 corresponding to the motor shaft position. It can be shown [59] that the kinetic and potential energies of the flexible-joint manipulator is defined as:

$$KE = \frac{1}{2}\dot{q}_1 D(q_1)\dot{q}_1 + \frac{1}{2}\dot{q}_2 J\dot{q}_2 \tag{B.8}$$

$$VE = V_1(q_1) + V_2(q_1, q_2), \tag{B.9}$$

where $D(q_1)$ and $V_1(q_1)$ are the inertia matrix and potential energy of the rigid manipulator as derived in Section B.2.1, respectively, J is a diagonal matrix corresponding to motor inertia:

$$J = diag\{\frac{1}{r_1^2}J_1, \ldots, \frac{1}{r_n^2}J_n\},$$

where, J_i is the inertia and r_i is the gear ratio of the ith rotor and V_2 is the elastic potential energy of the joint:

$$V_2(q_1, q_2) = \int \tau_s.dq$$

$$\tau_s = K(q_2 - q_1) + \beta(q_1, \dot{q}_1, q_2, \dot{q}_2), \tag{B.10}$$

where $K \in \Re^{n\times n}$ is the diagonal positive-definite stiffness matrix of the rotational spring including the joint spring constant, and $\beta(q_1,\dot{q}_1,q_2,\dot{q}_2)$ represents an unknown nonlinear force regarded as a combination of a nonlinear spring and friction at the output shafts of the manipulator. It should be noted that the reaction torque τ_s cannot be modeled precisely and is assumed to be unknown for observer design.

Moreover, another important source of nonrigid body forces is the friction. The force due to the friction can be quite large and in some cases up to 25% of the required torque for moving the manipulator in typical situation [135]. To make the dynamic equations reflecting the reality of physical device, it is important to model the frictions. A simple linear model for friction is the **viscous friction** Therefore, the torque, τ in Equation (B.6) can be modified as $\tau - \tau_{friction}$ for flexible-joint manipulators, where

$$\tau_{friction} = B[\dot{q}_1 \;\; \dot{q}_2]^T, \tag{B.11}$$

and $B = diag\{B_1, B_2\}$, $B_1 \in \Re^{n\times n}$ and $B_2 \in \Re^{n\times n}$ are the viscous damping matrices at the output and input shafts, respectively.

Thus, by considering (B.6), (B.8), (B.9), (B.10), and (B.11), the dynamics of flexible-joint manipulators obtained from the Euler-Lagrange equations can be described as follows:

$$\begin{aligned} D_l(q_1)\ddot{q}_1 + C_1(q_1,\dot{q}_1) + g(q_1) + B_1\dot{q}_1 &= \tau_s \\ J\ddot{q}_2 + \tau_s + B_2\dot{q}_2 &= \tau. \end{aligned} \tag{B.12}$$

Appendix C
Neural Network Learning Rules for Theorem 6.1

In this Appendix, the details for the weights updating mechanisms for the two neural networks prpoposed for FDI scheme in Theorem 6.1 are obtained. The learning rules given in (6.12)-(6.15) are based on the backpropagation algorithm. Hence, to implement (6.12)-(6.15), it is necessary to find $(\frac{\partial J}{\partial \hat{W}_1})$, $(\frac{\partial J}{\partial \hat{V}_1})$, $(\frac{\partial J}{\partial \hat{W}_2})$, and $(\frac{\partial J}{\partial \hat{V}_2})$.

Let us define

$$
\begin{aligned}
net_{\hat{v}_1} &= \hat{V}_1 \hat{\bar{x}} && \text{(C.1)}\\
net_{\hat{w}_1} &= \hat{W}_1 S_1 && \text{(C.2)}\\
net_{\hat{v}_2} &= \hat{V}_2 \hat{\bar{x}} && \text{(C.3)}\\
net_{\hat{w}_2} &= \hat{W}_2 S_3. && \text{(C.4)}
\end{aligned}
$$

Therefore, $\frac{\partial J}{\partial \hat{W}_1}$, $\frac{\partial J}{\partial \hat{V}_1}$, $\frac{\partial J}{\partial \hat{W}_2}$ and $\frac{\partial J}{\partial \hat{V}_2}$ can be computed as

$$
\begin{aligned}
\frac{\partial J}{\partial \hat{W}_1} &= \frac{\partial J}{\partial net_{\hat{w}_1}} \frac{\partial net_{\hat{w}_1}}{\partial \hat{W}_1}\\
\frac{\partial J}{\partial \hat{V}_1} &= \frac{\partial J}{\partial net_{\hat{v}_1}} \frac{\partial net_{\hat{v}_1}}{\partial \hat{V}_1}\\
\frac{\partial J}{\partial \hat{W}_2} &= \frac{\partial J}{\partial net_{\hat{w}_2}} \frac{\partial net_{\hat{w}_2}}{\partial \hat{W}_2}\\
\frac{\partial J}{\partial \hat{V}_2} &= \frac{\partial J}{\partial net_{\hat{v}_2}} \frac{\partial net_{\hat{v}_2}}{\partial \hat{V}_2}. && \text{(C.5)}
\end{aligned}
$$

We have

$$\frac{\partial J}{\partial net_{\hat{w}_1}} = \frac{\partial J}{\partial \tilde{y}} \frac{\partial \tilde{y}}{\partial \hat{x}} \frac{\partial \hat{x}}{\partial net_{\hat{w}_1}} = -\tilde{y}^T C \frac{\partial \hat{x}}{\partial net_{\hat{w}_1}}$$
$$\frac{\partial J}{\partial net_{\hat{v}_1}} = \frac{\partial J}{\partial \tilde{y}} \frac{\partial \tilde{y}}{\partial \hat{x}} \frac{\partial \hat{x}}{\partial net_{\hat{v}_1}} = -\tilde{y}^T C \frac{\partial \hat{x}}{\partial net_{\hat{v}_1}}$$
$$\frac{\partial J}{\partial net_{\hat{w}_2}} = \frac{\partial J}{\partial \hat{y}} \frac{\partial \hat{y}}{\partial net_{\hat{w}_2}} = -\tilde{y}^T \frac{\partial \hat{y}}{\partial net_{\hat{w}_2}}$$
$$\frac{\partial J}{\partial net_{\hat{v}_2}} = \frac{\partial J}{\partial \hat{y}} \frac{\partial \hat{y}}{\partial net_{\hat{v}_2}} = -\tilde{y}^T \frac{\partial \hat{y}}{\partial net_{\hat{v}_2}}, \tag{C.6}$$

and

$$\frac{\partial net_{\hat{w}_1}}{\partial \hat{W}_1} = S_1^T \tag{C.7}$$
$$\frac{\partial net_{\hat{v}_1}}{\partial \hat{V}_1} = \hat{\bar{x}}^T \tag{C.8}$$
$$\frac{\partial net_{\hat{w}_2}}{\partial \hat{W}_2} = S_3^T \tag{C.9}$$
$$\frac{\partial net_{\hat{v}_2}}{\partial \hat{V}_2} = \hat{\bar{x}}^T. \tag{C.10}$$

Now, by using (6.8) and the definitions of $net_{\hat{v}_1}$ and $net_{\hat{w}_1}$ in (C.1) and (C.2), respectively, the static approximation of the gradients $\frac{\partial \hat{x}(t)}{\partial net_{\hat{w}_1}}$ and $\frac{\partial \hat{x}(t)}{\partial net_{\hat{v}_1}}$ may be obtained as

$$\frac{\partial \hat{x}(t)}{\partial net_{\hat{w}_1}} = -J_0^{-1} \tag{C.11}$$
$$\frac{\partial \hat{x}(t)}{\partial net_{\hat{v}_1}} = -J_0^{-1} \hat{W}_1 (I - \Lambda(\hat{V}_1 \hat{\bar{x}})). \tag{C.12}$$

Moreover, the true value of the gradients $\frac{\partial \hat{y}(t)}{\partial net_{\hat{w}_2}}$ and $\frac{\partial \hat{y}(t)}{\partial net_{\hat{v}_2}}$ may be obtained from (6.8) and the definitions of $net_{\hat{v}_2}$ and $net_{\hat{w}_2}$ in (C.3) and (C.4), respectively, that is:

$$\frac{\partial \hat{y}(t)}{\partial net_{\hat{w}_2}} = I \tag{C.13}$$
$$\frac{\partial \hat{y}(t)}{\partial net_{\hat{v}_1}} = \hat{W}_2 (I - \Lambda(\hat{V}_2 \hat{\bar{x}})), \tag{C.14}$$

where

$$\Lambda(\hat{V}_j \hat{\bar{x}}) = diag\{\sigma_i^2(\hat{V}_j^i \hat{\bar{x}})\},\ j = 1,2,\ i = 1,2,\ldots,m, \tag{C.15}$$

and $\hat{V}_j^i$ denotes the ith row of the weight matrix $\hat{V}_j$. Now, let

$$S_2 = I - \Lambda(\hat{V}_1 \hat{\bar{x}})$$
$$S_4 = I - \Lambda(\hat{V}_2 \hat{\bar{x}}). \tag{C.16}$$

Note that S_2 and S_4 are in fact the derivatives of the tangent hyperbolic function which are also known to be bounded. Next, the gradients $\frac{\partial \hat{y}(t)}{\partial net_{\hat{w}_2}}$ and $\frac{\partial \hat{y}(t)}{\partial net_{\hat{v}_2}}$ given above can be multiplied by $-A_c^{-1}$, where A_c is a Hurwitz matrix with appropriate dimensions. This modification can give us a better control on the speed of convergence. Note that the sign of the gradients are not changed by this operation since A_c is a Hurwitz matrix, hence $-A_c^{-1}$ is a positive definite matrix. Consequently, we have:

$$\frac{\partial \hat{y}(t)}{\partial net_{\hat{w}_2}} \simeq -A_c^{-1} \tag{C.17}$$

$$\frac{\partial \hat{y}(t)}{\partial net_{\hat{v}_1}} \simeq -A_c^{-1} \hat{W}_2 (I - \Lambda(\hat{V}_2 \hat{\bar{x}})). \tag{C.18}$$

Now, by using equations (C.5)-(C.18), the learning rules (6.12)-(6.15) may be written as

$$\dot{\hat{W}}_1 = -\eta_1 (\tilde{y}^T C J_0^{-1})^T S_1^T - \rho_1 \| \tilde{y} \| \hat{W}_1 \tag{C.19}$$

$$\dot{\hat{V}}_1 = -\eta_2 (\tilde{y}^T C J_0^{-1} \hat{W}_1 S_2)^T \hat{\bar{x}}^T - \rho_2 \| \tilde{y} \| \hat{V}_1 \tag{C.20}$$

$$\dot{\hat{W}}_2 = -\eta_3 (\tilde{y}^T A_c^{-1})^T S_3^T - \rho_3 \| \tilde{y} \| \hat{W}_2 \tag{C.21}$$

$$\dot{\hat{V}}_2 = -\eta_4 (\tilde{y}^T A_c^{-1} \hat{W}_2 S_4)^T \hat{\bar{x}}^T - \rho_4 \| \tilde{y} \| \hat{V}_2.$$

Appendix D
Stability Conditions of Theorem 6.1-Part 2

The objective of this appendix is to find conditions on state estimation error $\tilde{x}$ and neural network weights error $\tilde{W}_2$ proposed in Theorem 6.1 that result in negative definiteness of the time derivative of the Lyapunov function candidate L. Towards this end, let us consider F as defined in (6.33) and rearranged here:

$$F = -\alpha_1 \|\tilde{x}\|^2 + \alpha_2 \|\tilde{x}\| - \alpha_{10} \|\tilde{W}_2\|^2 + \alpha_{11} \|\tilde{W}_2\|^2 - \alpha_{12} \|\tilde{W}_2\|^3 + F_1 + F_2 + F_3, \tag{D.1}$$

where

$$\begin{aligned} F_1 &= \|\tilde{x}\| \left(\alpha_3 \|\tilde{W}_1\| - \alpha_5 \|\tilde{W}_1\|^2 + \alpha_7 \|\tilde{W}_2\| - \alpha_9 \|\tilde{W}_2\|^2\right) \\ F_2 &= \|\tilde{W}_2\| \left(\alpha_{13} \|\tilde{W}_1\| - \alpha_{14} \|\tilde{W}_1\|^2 + \alpha_8\right) \\ F_3 &= \alpha_4 \|\tilde{W}_1\| - \alpha_6 \|\tilde{W}_1\|^2. \end{aligned} \tag{D.2}$$

Next, by completing the squares involving $\tilde{W}_1$ and $\tilde{W}_2$, we get

$$\begin{aligned} F_1 &= \|\tilde{x}\| \left(-\left(\sqrt{\alpha_5} \|\tilde{W}_1\| - \frac{\alpha_3}{2\sqrt{\alpha_5}}\right)^2 + \frac{\alpha_3^2}{4\alpha_5} - \left(\sqrt{\alpha_9} \|\tilde{W}_2\| - \frac{\alpha_7}{2\sqrt{\alpha_9}}\right)^2 + \frac{\alpha_7^2}{4\alpha_9} \right) \\ \therefore F_1 &\leq \beta_1 \|\tilde{x}\|, \ \beta_1 = \frac{\alpha_3^2}{4\alpha_5} + \frac{\alpha_7^2}{4\alpha_9} \end{aligned} \tag{D.3}$$

$$\begin{aligned} F_2 &= \|\tilde{W}_2\| \left(-\left(\sqrt{\alpha_{14}} \|\tilde{W}_1\| - \frac{\alpha_{13}}{2\sqrt{\alpha_{14}}}\right)^2 + \alpha_8 + \frac{\alpha_{13}^2}{4\alpha_{14}} \right) \\ \therefore F_2 &\leq \beta_2 \|\tilde{W}_2\|, \ \beta_2 = \alpha_8 + \frac{\alpha_{13}^2}{4\alpha_{14}} \end{aligned} \tag{D.4}$$

$$\begin{aligned} F_3 &= -\left(\sqrt{\alpha_6} \|\tilde{W}_1\| - \frac{\alpha_4}{2\sqrt{\alpha_6}}\right)^2 + \frac{\alpha_4^2}{4\alpha_6} \\ \therefore F_3 &\leq \frac{\alpha_4^2}{4\alpha_6}. \end{aligned} \tag{D.5}$$

Now, by using (D.3)-(D.5), we can write:

$$F \leq -\alpha_1 \|\tilde{x}\|^2 + \alpha_2 \|\tilde{x}\| + \beta_1 \|\tilde{x}\| + \beta_2 \|\tilde{W}_2\| + \frac{\alpha_4^2}{4\alpha_6} - \alpha_{10} \|\tilde{W}_2\|^2 + \alpha_{11} \|\tilde{W}_2\|^2 - \alpha_{12} \|\tilde{W}_2\|^3. \tag{D.6}$$

Next, by using some algebraic manipulations, it may be shown that:

$$-\alpha_{12} \|\tilde{W}_2\|^3 + \alpha_{11} \|\tilde{W}_2\|^2 \leq -\alpha_{12} \|\tilde{W}_2\|^2 + \beta_3, \tag{D.7}$$

where

$$\beta_3 = \frac{4}{27} \frac{(\alpha_{11} + \alpha_{12})^3}{\alpha_{12}^2}. \tag{D.8}$$

Hence, by substituting (D.7) in (D.6), we get

$$F \leq -\alpha_1 \|\tilde{x}\|^2 + \alpha_2 \|\tilde{x}\| + \beta_1 \|\tilde{x}\| + \beta_2 \|\tilde{W}_2\| + \frac{\alpha_4^2}{4\alpha_6} - (\alpha_{10} + \alpha_{12}) \|\tilde{W}_2\|^2 + \beta_3. \tag{D.9}$$

Completing the squares involving $\|\tilde{W}_2\|$ in the above equation results in

$$F \leq -\alpha_1 \|\tilde{x}\|^2 + (\alpha_2 + \beta_1) \|\tilde{x}\| + \beta_4 \tag{D.10}$$

$$\beta_4 = \beta_3 + \frac{\alpha_4^2}{4\alpha_6} + \frac{\beta_2^2}{4(\alpha_{10} + \alpha_{12})}.$$

Similarly, completing the squares involving $\|\tilde{x}\|$ in (D.9) yields

$$F \leq -(\alpha_{10} + \alpha_{12}) \|\tilde{W}_2\|^2 + \beta_2 \|\tilde{W}_2\| + \beta_5 \tag{D.11}$$

$$\beta_5 = \beta_3 + \frac{\alpha_4^2}{4\alpha_6} + \frac{(\alpha_2 + \beta_1)^2}{4\alpha_1}.$$

The right hand side of the above inequalities, i.e., (D.10) and (D.11), represent second order algebraic equations. Consequently, provided that $\lambda_{min}(Q) > 2l_g \|P\|$, F (and hence $\dot{L}$) is negative semi-definite if

$$\|\tilde{x}\| \geq x^0 = \frac{(\alpha_2 + \beta_1) + \sqrt{(\alpha_2 + \beta_1)^2 + 4\beta_4\alpha_1}}{2\alpha_1}, \tag{D.12}$$

Or

$$\|\tilde{W}_2\| \geq w_2^0 = \frac{\beta_2 + \sqrt{\beta_2^2 + 4\beta_5(\alpha_{10} + \alpha_{12})}}{2(\alpha_{10} + \alpha_{12})}.$$

References

1. H. K. Khalil, *Nonlinear Systems*. Prentice Hall, third ed., 2002.
2. K. S. Narendra and J. H. Taylor, *Frequency Domain Criteria for Absolute Stability*. Academic Press, New York, 1973.
3. M. Spong and M. Vidyasagar, *Robot Dynamics and Control*. McGraw Hill, 1989.
4. E. Bullinger and F. Allgower, "An adaptive high-gain observer for nonlinear systems," in *36th IEEE Conference on Decision and Control, San Diego, California*, pp. 4348–4353, Dec. 1997.
5. Canadian Space Agency, "Canadarm2." `http://www.asc-csa.gc.ca/asc/app/gallery/results2.asp?image_id=Canadarm2-007`, access Dec. 2008.
6. S. A. Bogh and M. Blanke, "Fault-tolerant control - a case study of the ørsted satellite," *IEE Colloquium on Fault Diagnosis in Process Systems*, pp. 11/1–11/13, Apr. 1997.
7. L. Anthony, "The basics of harmonic drive gearing," *Gear Product News*, pp. 32–36, Apr. 2006.
8. Harmonic Drive Systems Inc., "Principles of harmonic drive." `http://www.hds.co.jp/HDS_hp_english/english/index.html`, access Dec. 2008.
9. K. Schugerl, "Progress in monitoring, modeling and control of bioprocesses during the last 20 years," *Journal of Biotechnology,* vol. 85, no. 2, pp. 149–173, Feb. 2001.
10. M. Soroush, "State and parameter estimations and their applications in process control," *Computers and Chemical Engineering* , vol. 23, no. 2, pp. 229–245, 1998.
11. I. E. Potter and M. C. Sunman , "Thresholdless redundancy management with arrays of skewed instruments," *Integrity in Electronic Flight Control Systems AGARDOGRAPH-224*, pp. 15–11–15–25, 1977.
12. X. C. Lou, A. S. Willsky and G. L. Verghese, "Optimally robust redundancy relations for failure detection in uncertain systems," *Automatica* vol. 22, no. 3, pp. 333–344, May 1986.
13. D. N. Wilbers, and J. L. Speyer, "Detection filters for aircraft sensor and actuator faults," in *'89th IEEE Int. Conference on Control and Applications, Jerusalem*, pp. 81–86, 1989.
14. R. Isermann, "Process fault diagnosis based on modeling and estimation methods-a survey," *Automatica,* vol. 20, pp. 387–404, 1984.
15. P. M. Frank, "Fault diagnosis in dynamic systems using analytical and knowledge-based redundancy-a survey and some new results," *Automatica,* vol. 26, no. 3, pp. 459 –474, May 1990.
16. A. T. Vemuri, M. M. Polycarpou, and S. A. Diakourtis, "Neural network based fault detection in robotic manipulators," *IEEE Transactions on Robotics and Automation* vol. 14, no. 2, pp. 342–348, Apr. 1998.
17. R. Mehra, S. Seereeram, D. Bayard, and F. Hadaegh, "Adaptive Kalman filtering, failure detection and identification for spacecraft attitude estimation," in *4th IEEE Conference on Control Applications*, pp. 176–181, 1995.
18. E. Sobhani, K. Khorasani and S. Tafazoli, "Dynamic neural network-based estimator for fault diagnosis in reaction wheel actuator of satellite attitude control system," in *Int. Joint Conference on Neural Networks*, pp. 2347–2352, 2005.
19. X. D. Zhang, T. Parisini and M. M. Polycarpou, "Sensor bias fault isolation in a class of nonlinear systems," *IEEE Transactions on Automatic Control,* vol. 50, no. 3, pp. 370–376, Mar. 2005.
20. M. McIntyre, W. E. Dixon, D. M. Dawson, and I. D. Walker, "Fault identification for robot manipulators," *IEEE Transactions on Robotics,* vol. 21, no. 5, pp. 1028–1034, Oct. 2005.
21. D. G. Luenberger, "Observing the state of a linear systems," *IEEE Transactions Military Electronics* vol. 8, no. 2 , pp. 74–80, 1964.
22. P. Swerling, "A proposed stagewise differential correction procedure for satellite tracking and prediction," *Technical report P-1292 Rand Corporation*, Jan 1958.
23. R. E. Kalman, "A new approach to linear filtering and prediction problems," *Journal of Basic Engineering,* vol. 82, no. 1, pp. 35–45, 1960.
24. J.B. Burl, *Linear Optimal Control*. Addison-Wesley Longman, 1999.

25. L. Ljung, "Asymptotic behavior of the extended Kalman filter as a parameter estimator for linear systems," *IEEE Transactions on Automatic Control,* vol. 24, no. 1, pp. 36–50, Feb. 1979.
26. S. Haykin, *Kalman Filtering and Neural Networks.* John Wiley & Sons, 2001.
27. A. Tornambe, "Use of asymptotic observers having high gains in the state and parameter estimation," in *IEEE Conference on Decision and Control, Tampa, Florida*, pp. 1791–1794, 1989.
28. S. Nicosia and A. Tornambe, "High-gain observer in the state and parameter estimation of robots having elastic joints," *System and Control Letters,* vol. 13, pp. 331–337, 1989.
29. Q. Zhang and A. Xu, "Implicit adaptive observer for a class of nonlinear systems," in *American Control Conference*, pp. 1551–1556, Jun. 2001.
30. M. Farza, M. M. Saad., and L. Rossignol, "Observer design for a class of MIMO nonlinear systems," *Automatica* vol. 40, no. 1, pp. 135–143, Jan. 2004.
31. P. Krishnamurthy, F. Khorrami, and R. S. Chandra, "Global high-gain-based observer and backstepping controller for generalized output-feedback canonical form," *IEEE Transactions on Automatic Control,* vol. 48, no. 12, pp. 2277–2284, Dec. 2003.
32. J. J. E. Slotine and W. Li, *Applied Nonlinear Control.* Prentice-Hall Inc., 1991.
33. Y. Xiong and M. Saif, "Sliding mode observer for nonlinear uncertain systems," *IEEE Transactions on Automatic Control,* vol. 46, no. 12, pp. 2012–2017, Dec. 2001.
34. A. J. Koshkouei and A. S. I. Zinober, "Sliding mode state observation for non-linear systems," *International Journal of Control,* vol. 77, no. 2, pp. 117–127, Jan. 2004.
35. W. Yu, M.A. Marino and X. Li, "Observer-based neuro identifier," *IEE Proceedings of Control Theory and Applications,* vol. 147, no. 2, pp. 145–152, Mar. 2000.
36. J. Resendiz, W. Yu, and L. Fridman, "Two-stage neural observer for mechanical systems," *IEEE Transactions on Circuits and Systems II: express Brief,* vol. 55, no. 10, pp. 1076–1080, Oct. 2008.
37. J. Davila, L. Fridman, and A. Levant, "Second-order sliding-mode observer for mechanical systems," *IEEE Transactions on Automatic Control,* vol. 50, no. 11, pp. 1785–1789, Nov. 2005.
38. P. B. Goldsmith, B. A. Francis, and A. A. Goldenderg, "Stability of hybrid position/force control applied to manipulators with flexible joints," *International Journal of Robotics and Automation,* vol. 14, no. 4, pp. 146–159, 1999.
39. G. Cybenko, "Approximation by superposition of a sigmoidal function," *Mathematics of Control, Signals and Systems,* vol. 2, no. 4, pp. 303–314, Dec. 1989.
40. K. I. Funahashi, "On the approximation of realization of continuous mappings by neural networks," *Neural Networks,* vol. 2, pp. 183–192, 1989.
41. R. H. Nielsen, "Theory of the backpropagation neural network," in *International Joint Conference on Neural Networks*, pp. 593–605, 1989.
42. J. M. Zurada, *Introduction to Artificial Neural Systems.* West Publishing Company, 1992.
43. P. Werbos, "Generalization of backpropagation with application to recurrent gas market model," *Neural Networks* vol. 1, pp. 339–356, 1988.
44. F. Abdollahi, H. A. Talebi, and R. V. Patel, "A stable neural network-based observer with application to flexible-joint manipulators," *IEEE Transactions on Neural Networks* vol. 17, no. 1, pp. 118–129, Jan. 2006.
45. J. Y. Choi and J. A. Farrell, "Adaptive observer backstepping control using neural networks," *IEEE Transactions on Neural Networks,* vol. 12, no. 5, pp. 1103–1112, Sep. 2001.
46. V. Stepanyan and N. Hovakimyan, "Robust adaptive observer design for uncertain systems with bounded disturbances," *IEEE Transactions on Neural Networks,* vol. 18, no. 5, pp. 1392–1403, Sep. 2007.
47. M. S. Ahmed and S. H. Riyaz, "Dynamic observer-a neural net approach," *Intelligent and Fuzzy System* vol. 9, no. 1, pp. 113–127, Nov. 2000.
48. Y. G. Leu, W. Y. Wang, and T. T. Lee, "Observer-based direct adaptive fuzzy-neural control for nonaffine nonlinear systems," *IEEE Transactions on Neural Networks,* vol. 16, no. 4, pp. 853–863, Jul. 2005.

49. Y. H. Kim, F. L. Lewis, and C. T. Abdallah, "Nonlinear observer design using dynamic recurrent neural networks," in *35th IEEE Conference on Decision and Control, Kobe, Japan*, pp. 949–954, Dec. 1996.
50. R. Zhu, T. Chai and C. Shao, "Robust nonlinear adaptive observer design using dynamic recurrent neural networks," in *American Control Conference Albuquerque,New Mexico*, pp. 1096–1100, Jun. 1997.
51. A. S. Poznyak, E. N. Sanchez, O. Palma and W. Yu, "Output trajectory tracking using dynamic neural networks," in *39th IEEE Conference on Decision and Control*, pp. 889–894, Dec. 2000.
52. J. A. R. Vargas and E. M. Hemerly, "Robust neural adaptive observer for MIMO nonlinear systems," in *IEEE Conference on Systems, Man, and Cybernetics*, pp. 1084–1089, 1999.
53. J. A. R. Vargas and E. M. Hemerly, "Neural adaptive observer for general nonlinear systems," in *American Control Conference*, pp. 708–712, 2000.
54. H. A. Talebi, R. V. Patel and M. Wong, "A neural network based observer for flexible-joint manipulators," in *15th IFAC World Congress on Automatic Control, Barcelona, Spain*, Jul. 2002.
55. K. S. Narendra and K. Parthasarathy, "Identification and control of dynamical systems using neural networks," *IEEE Transactions on Neural Networks*, vol. 1, no. 1, pp. 4–27, Mar. 1990.
56. H. A. Talebi, R. V. Patel and H. Asmer, "Neural network based dynamic modeling of flexible-link manipulators with application to the SSRMS," *Journal of Robotic Systems*, vol. 17, no. 7 , pp. 385–401, May 2000.
57. M. W. Spong, "On the force control problem for flexible joint manipulators," *IEEE Transactions Automotic Control* vol. 34, no. 1, pp. 107–111, Jan 1989.
58. S. S. Ge and L. C. Woon, "Adaptive neural network control of flexible joint manipulators in constrained motion," *Transactions on the Institute of Measurement and Control,* vol. 20, no. 1, pp. 37–46, 1998.
59. M. W. Spong, "Modeling and control of elastic joint robots," *ASME Journal Dynamics Systems Measurement and Control,* vol. 109, pp. 310–319, Dec. 1987.
60. Y. Z. Chang and R. W. Daniel, "On the adaptive control of flexible joint robots," *Automatica,* vol. 28, no. 5, pp. 969–974, Sep. 1992.
61. V. Zeman, R. V. Patel, and K. Khorasani, "Control of a flexible-joint robot using neural networks," *IEEE Transactions on Control Systems Technology,* vol. 5, no. 4, pp. 453–462, Jul. 1997.
62. J.Y. Hung, "Control of industrial robots that have transmission elasticity," *IEEE Transactions on Industrial Electronics,* vol. 38, no. 6 , pp. 421–427, Dec. 1991.
63. R.A. Al-Ashoor, R. V. Patel, and K. Khorasani, "Robust adaptive controller design and stability analysis for flexible-joint manipulators," *IEEE Transactions on Systems, Man, and Cybernetics,* vol. 23, no. 2, pp. 589–602, Mar. 1993.
64. M. W. Spong, K. Khorasani, and P. V. Kokotovic, "An integral manifold approach to feedback control of flexible joint robots," *IEEE Journal of Robotics and Automation,* vol. 3, no. 4, pp. 291–301, Aug. 1987.
65. K. Khorasani, "Adaptive control of flexible-joint robots," *IEEE Transactions on Robotics and Automation,* vol. 8, no. 2, pp. 250–267, Apr. 1992.
66. M. Jankovic, "Observer based control for elastic joint robots," *IEEE Transactions Robotics and Automation,* vol. 11, no. 4 , pp. 618–623, 1995.
67. S. Nicosia, P. Tomei, and A. Tornambe, "A nonlinear observer for elastic robots," *IEEE Transactions Robotics and Automation,* vol. 4, no. 1 , pp. 45–52, 1988.
68. S. Nicosia, P. Tomei, and A. Tornambe, "An approximate observer for a class of nonlinear systems," *System and Control Letters,* vol. 12, no. 1, pp. 43–51, 1989.
69. P. Tomei, "An observer for flexible joint robots," *IEEE Transactions Automotic Control,* vol. 35, no. 6, pp. 739–743, Jun. 1990.
70. M. Jankovic, "Exponentially stable observer for elastic joint robots," in *31st IEEE Conference on Decision and Control*, pp. 323–324, 1992.

71. A. De Luca and B. Siciliano, "An asymptotically stable joint PD controller for robot arms with flexible links under gravity," in *31st IEEE Conference on Decision and Control*, pp. 325–326, 1992.
72. F. Abdollahi, H. A. Talebi, and R. V. Patel, "A stable neural network-based identification scheme for nonlinear systems," in *American Control Conference, Denver*, pp. 3590–3595, 2003.
73. W. Ren and P. R. Kumar, "Stochastic adaptive prediction and model reference control," *IEEE Transactions on Automatic Control,* vol. 39, no. 10, pp. 2047–2060, Oct. 1994.
74. F. L. Lewis, S. Jagannathan, and A. Yesildirek, *Neural Network Control of Robot Manipulators and Nonlinear Systems*. Taylor & Francis, 1999.
75. M. M. Polycarpou, "Stable adaptive neural control scheme for nonlinear systems," *IEEE Transactions on Automatic Control,* vol. 41, no. 3, pp. 447–451, Mar. 1996.
76. N. Sadegh, "A perceptron network for functional identification and control of nonlinear systems," *IEEE Transactions on Neural Networks,* vol. 4, no. 6, pp. 982–988, Nov. 1993.
77. M. M. Polycarpou and P. A. Ioannou, "Modeling, identification and stable adaptive control of continuous-time nonlinear dynamical system using neural networks," in *American Control Conference*, pp. 36–40, 1992.
78. C. C. Lee, and C. C. Teng, "Identification and control of dynamic system using recurrent fuzzy neural networks," *IEEE Transaction on Fuzzy Systems,* vol. 8, no. 4, pp. 349–366, Aug. 2000.
79. J. Mahdavi, M.R. Nasiri, A. Agah, and A. Emadi, "Application of neural networks and state-space average to DC/DC PWM converters in sliding-mode operation," *IEEE/ASME Transaction on Mechatronics,* vol. 10, no. 1, pp. 60–67, Feb. 2005.
80. N. Hovakimyan, A. J. Calise and V. K. Madyastha, "An adaptive observer design methodology for bounded nonlinear processes," in *41st IEEE Conference on Decision and Control*, pp. 4700–4705, Dec. 2002.
81. K. S. Narendra and A. N. Annaswamy, *Stable Adaptive Systems*. Prentice Hall, Enlgewood Cliffs, NJ, 1989.
82. K. H. Low and M. Vidyasagar, "A lagrangian formulation of the dynamic model for flexible manipulator systems," *Journal of Dynamic Systems, Measurement, and Control,* vol. 110, no. 2, pp. 175–181, Jun. 1988.
83. A. D. Luca and B. Siciliano, "Explicit dynamic modeling of a planar two-link flexible manipulator," in *29th IEEE Conference on Decision and Control*, pp. 528–530, 1990.
84. Y. Huang and C. S. G. Lee, "Generalization of newton-euler formulation of dynamic equations to nonrigid manipulators," in *American Conference Control*, pp. 72–77, Jun. 1987.
85. A.G. Kelkar, S.M. Joshi, and T.E. Alberts, "Globally stabilizing controllers for flexible multibody systems," in *31st IEEE Conference on Decision and Control*, pp. 2856–2859, 1992.
86. Hassan K. Khalil, *Nonlinear Systems*. Prentice Hall, 2001.
87. Wikipedia , "Canadarm2." http://en.wikipedia.org/wiki/SSRMS, access Dec. 2008.
88. A. Sharon and D. Hardt, "Enhancement of robot accuracy using endpoint feedback and a macro-micro manipulator system," in *American Control Conference, San Diego, CA*, pp. 1836–1842, 1984.
89. D. N. Nenchev, K. Yoshida, P. Vichitkulsawa, M. Uchiyama, "Recreation null-space control of flexible structure mounted manipulator systems," *IEEE Transactions on Robotic and Automation* vol. 15, no. 6, pp. 1011–1023, Dec. 1999.
90. J. F. Jansen, B. L. Burks, S. M. Babcock, R. L. Kress, and W. R. Hamel, "Long-reach manipulation for waste storage tank remediation," in *American Society of Mechanical Engineers (ASME) annual winter meeting, Atlanta, GA*, Dec. 1991.
91. D. S. Kwon, D. H. Hwang, S. M. Babcock, and B. L. Burks, "Input shaping filter methods for the control of structurally flexible, long-reach manipulators," in *IEEE International Conference on Robotics and Automation, San Diego, CA*, pp. 3259–3264, May 1994.
92. M. A. Torres and S. Dubowsky, "Path-planning in elastically constrained space manipulator systems," in *IEEE International Conference on Robotics and Automation, Atlanta, GA*, pp. 812–817, May 1993.

93. C. Vallancourt and C. M. Gosselin, "Compensating for the structural flexibility of the ssrms with the spdm," in *2nd Workshop Robot Space, Canadian Space Agency, Montreal, PQ, Canada*, Jul. 1994.
94. A. Mannani and H. A. Talebi, "Implementing of a fuzzy Lyapunov-based control strategy for a Macro-Micro manipulator," in *European Control Conference*, Sep. 2003.
95. X. P. Cheng and R. V. Patel, "Neural network based tracking control of a flexible macro-micro manipulator system," *Neural Networks*, vol. 16, no. 2, pp. 271–286, Sep. 2003.
96. D. P. Magee and W. J. Book, "Filtering macro manipulator wrist commands to prevent flexible base motion," in *American Control Conference, Seattle, WA*, pp. 1836–1842, 1995.
97. M. Moallem and R. V. Patel, "A vibration control strategy for a boom-mounted manipulator system for high-speed positioning," in *IEEE International Conference on Intelligent Robots and Systems, Takamatsu, Japan*, pp. 299–304, 1999.
98. M. R. Kermani, R. V. Patel, and M. Moallem, "Flexure control using piezostack actuators: Design and implementation," *IEEE/ASME Transactions on Mechatronics*, vol. 10, no. 2, pp. 181–188, Apr. 2005.
99. M. L. Visinsky, J. R. Cavallaro, and I. D. Walker, "Expert system framework for fault detection and fault tolerance in robotics," *Computers and Electrical Engineering*, vol. 20, no. 5, pp. 65–75, Sep. 1994.
100. W. Bucken and B. Freyermuth, "Fault detection in industrial processes by statistical methods in the example of an industrial robots," *ITG-Fachberichte*, pp. 123–136, 1991.
101. B. Freyermuth, "An approach to model based fault diagnosis of industrial robots," in *IEEE International Conference on Robotics and Automation*, pp. 1350–1356, 1991.
102. T. W. S. Chow and H. Z. Tan, "HOS-based nonparametric and parametric methodologies for machine fault detection," *IEEE Transactions on Industrial Electronics*, vol. 47, no. 5, pp. 1051–1059, Oct. 2000.
103. R. Tinos, M. H. Terra, and M. Bergerman, "Fault detection and isolation in cooperative manipulators via artificial neural networks," in *IEEE International Conference on Control Applications*, pp. 492–497, 2001.
104. M. H. Terra and R. Tinos, "Fault detection and isolation in robotic systems via artificial neural networks," in *IEEE Conference on Decision and Control*, pp. 1605–1610, 1998.
105. J. M. Naughton, Y. C. Chen, and J. Jiang, "A neural network application to fault diagnosis for robotics manipulators," in *IEEE International Conference on Control Applications*, pp. 988–993, 1996.
106. H. Schneider and P. M. Frank, "Observer-based supervision and fault detection in robots using nonlinear and fuzzy logic residual evaluation," *IEEE Transactions on Control System Technology*, vol. 4 no. 3, pp. 274–282, May 1996.
107. S. Wu and T. W. S. Chow, "Induction machine fault detection using SOM-based RBF neural networks," *IEEE Transactions on Industrial Electronics*, vol. 51, no. 1, pp. 183–194, Feb. 2004.
108. S. C. Tan and C. P. Lim, "Application of an adaptive neural network with symbolic rule extraction to fault detection and diagnosis in a power generation plant," *IEEE Transactions on Energy Conversion*, vol. 19, no. 2, pp. 369–377, Jun. 2004.
109. T. Bak, R. Wisniewski and M. Blanke, "Autonomous attitude determination and control system for the ørsted satellite," in *IEEE Aerospace Applications Conference*, pp. 173–186, 1996.
110. S. Murugesan and P. S. Goel, "A scheme for fault tolerance in earth sensors," *IEEE Transactions on Aerospace and Electronic Systems*, vol. 25, no. 1, pp. 21–30, Jan. 1989.
111. I. A. Al-Zyoud and K. Khorasani, "Detection of actuator faults using dynamic neural network for the attitude control subsystem of a satellite," in *Int. Joint Conference on Neural Networks*, pp. 1746–1751, 2005.
112. W. E. Dixon, I. D. Walker, D. M. Dawson, and J. P. Hartranft, "Fault detection for robot manipulators with parametric uncertainty: A prediction-error-based approach," *IEEE Transactions on Robotics and Automation*, vol. 16, no. 6, pp. 689–699, Dec. 2000.
113. F. Bacconi, D. Angeli and E. Mosca, "Attitude control of asymmetric spacecrafts subject to actuator failures," in *IEEE Conference on Control Applications*, pp. 474–479, 2003.

114. Q. Wu and M. Saif, "Neural adaptive observer based fault detection and identification for satellite attitude control systems," in *American Control Conference*, pp. 1054–1059, 2005.
115. Q. Wu and M. Saif, "Robust fault diagnosis for satellite attitude systems using neural state space models," in *IEEE International Conference on Systems, Man and Cybernetics*, pp. 1955–1960, 2005.
116. Q. Wu and M. Saif, "Robust fault diagnosis for a satellite system using a neural sliding mode observer," in *IEEE Conference on Decision and Control*, pp. 7668–7673, 2005.
117. Q. Wu and M. Saif, "Robust fault diagnosis for a satellite large attitude system using an iterative neuron PID (INPID) observer," in *American Control Conference*, pp. 5710–5714, 2006.
118. J. E. Slotine and M. D. Di Benedetto, "Hamiltonian adaptive control of spacecraft," *IEEE Transactions on Automatic Control,* vol. 35, no. 7, pp. 848–852, Jul. 1990.
119. L. Show, J.C. Juang and C.D. Yang, "Nonlinear H_∞ robust control for satellite large angle attitude maneuvers," in *American Control Conference*, pp. 1357–1362, 2001.
120. M. V. A. Corp., *TRC-041 User's Manual.* Mark V Automation Corp., North Oaks, MN, USA, 2000.
121. I. Wind River Systems, *VxWorks Programmer's Guide 6.6.* Wind River Systems, Alameda, CA, 94501-1153, USA, 2007.
122. *Real-Time Workshop User's Guide.* MathWorks, Natick, MA, USA, 2000.
123. B. Armstrong, O. Khatib and J. Burdick, "The explicit dynamic model and inertial parameters of the PUMA 560 arm," in *Proc. IEEE Int. Conference Robotics and Automation*, pp. 510–518, 1986.
124. H. A. Talebi, K. Khorasani, and S. Tafazoli, "A recurrent neural network-based sensor and actuator fault detection and isolation for nonlinear systems with application to the satellite's attitude control subsystem," *IEEE Transaction on Neural Networks,* vol. 20, no. 1, pp. 45–60, Jan. 2009.
125. H. A. Talebi and R. V. Patel, "An intelligent fault detection and recovery scheme for reaction wheel actuator of satellite attitude control systems," in *IEEE Conference on Control Applications*, 2006.
126. I. A. Al-Zyoud and K. Khorasani, "Neural network-based actuator fault diagnosis for attitude control subsystem of an unmanned space vehicle," in *Int. Joint Conference on Neural Networks*, pp. 3686 – 3693, 2006.
127. Z. Q. Li, L. Ma and K. Khorasani, "A dynamic neural network-based reaction wheel fault diagnosis for satellites," in *World Congress on Computational Intelligence (WCCI'06)*, 2006.
128. C. A. Lindley, *Top-Level Design Document for Project AUSTRALIS-1.* Australian Space Research Institute Ltd, access Dec 2008.
129. F. Martel, K. Parimal, and M. Psiaki, "Active magnetic control system for gradient stabilized spacecraft," in *Proceeding of Annual AIAA/USU Conference on Small Satellites*, Sep. 1988.
130. J. E. Sedlak, "Improved spacecraft attitude filter using a sequentially correlated magnetometer noise model," in *Digital Avionics Systems Conference (DASC'97)*, pp. 8.4–9 – 8.4–16, 1997.
131. Y. V. Kim, K. J. Di Filippo, and A. Ng, "On the calibration of satellite on-board magnetometer," in *The Fourth International Conference on Control and Automation*, pp. 947–951, 2003.
132. R. Wisniewski, *Satellite Attitude Control Using Only Electromagnetic Actuation.* PhD thesis, Department of Control Engineering, Aalborg University, Aalborg, Denmark, 1996.
133. Wikipedia , "Harmonic drive." `http://en.wikipedia.org/wiki/Harmonic_drive`, access Dec. 2008.
134. H. Goldstein, *Classical Mechanics.* Addison-Wesley, Reading, MA, 1981.
135. J. J. Craig, *Introduction to Robotics.* Pearson Prentice Hall, Third Ed., 1995.

Index